2017年地质院校特色英语精品教材项目（编号：ZL201721）资助
《地质英语阅读教程》MOOC课程建设项目（编号：ZL201834）资助
2017年大学英语课程建设研究与实践项目（编号：ZL201701）资助
2017年教学研究立项评审结果A类-29项目资助

赖旭龙 董元兴 刘 芳 赖小春 丛书主编

ESP
地学英语阅读系列教材

地学英语阅读教程
Comprehensive English Reading for Geosciences

赖小春 汪卫红 主 编

王晓婧 付 蕾 曾艳霞 胡冬梅 刘雪莲 副主编

中国地质大学出版社
CHINA UNIVERSITY OF GEOSCIENCES PRESS

《地学英语阅读系列教材》

编撰委员会

主　编	赖旭龙	董元兴	刘　芳	赖小春	
顾　问	唐辉明	刘勇胜	周爱国	马昌前	殷坤龙
	胡祥云	李建威	马　腾	章军锋	牟　林

编　委（按姓氏笔画排序）

王　伟	王四海	王国念	王晓婧	卢　云	付　蕾
冯　迪	任利民	刘春华	刘晓琴	刘倩倩	刘雪莲
刘　敏	江　翠	严　瑾	杨红燕	肖　珊	何　霜
汪卫红	张红燕	张伶俐	张英贤	张　莉	张基得
陈　俐	陈晓斌	周宏图	周诗雨	赵　妍	胡冬梅
胡志红	姚夏晶	秦　屹	黄曼丽	彭　林	葛亚非
曾艳霞	蓝希君	赖小春			

《地学英语阅读教程》

主　编　赖小春　汪卫红
副主编　王晓婧　付　蕾　曾艳霞　胡冬梅　刘雪莲

Preface

I welcome CUG Press's publication of a volume of Earth Science texts in English accompanied by readings by native speakers, and I urge you to listen, read texts aloud, record them and listen and compare yourself.

English is the international language of Earth Sciences used to publish results in international journals and exchange ideas at meetings all round the world. Nowadays Earth Science graduates from Chinese universities including CUG have English language skill in reading, but are behind non-native speakers from Europe, the Americas and the Indian sub-continent in listening, speaking and writing. I have seen an immense improvement at CUG over the last 30 years but my experience in classes, seminars, meetings, thesis defences and editing the English of papers submitted to CUG's *Journal of Earth Sciences* convinces me that further improvement is needed.

Search for the meaning when you read silently. Begin by reading quickly ("skimming") to get the general meaning of a paragraph and write down in English what you judge to be the main meaning. Then read again more carefully. The structure of English sentences is different from Chinese because a writer will begin with the main subject and give evidence in the sentence complement afterwards. For example, a native speaker might write "a *P-T* evolution diagram of the Granites has been drawn using calculated crystallization temperatures and temperature changes during magma evolution", rather than "based on the calculated temperatures of the Granites and temperature changes during magma evolution, a *P-T* evolution diagram of Granite can be drawn".

Spoken English uses voice tones differently from Chinese. Listen to the recordings of the texts for examples. You should already know that a rising tone at the end of a sentence signals a question, "Is this the way to the rock mechanics *láb*?" Voice tone is often raised for emphasis and will help you to recognise key words and phrases as you listen to a talk. "There was a *major change* in Earth's tectonics about 2.1 billion years ago." Falling tones indicate disagreement, "Professor, I think your theory is *wròng*."

Written English derives from speech and you should use every opportunity you can to listen and speak. Local spoken pronunciation varies but at scientific meetings the English is almost always correct, so if you don't understand a native speaker you need to improve your listening to include several varieties. (Be tolerant of non-Chinese non-native English speakers who face the same difficulties as you.) Chinese learners of English worry about differences between American and British English but they are small compared with the varieties of English spoken in different parts of the United States and the British Isles, and of course there are also distinctive types of English in Canada, Australia, New Zealand, and South Africa. Millions of people in India, Pakistan, Sri Lanka, Bangladesh, Singapore, and some African countries speak English as their first language but have never used it to converse with a British or an American person! We have included examples of non-standard English pronunciation among our readings.

Relax while you are learning and enjoy the texts and readings while you learn from them. Don't worry about making mistakes because your teachers will help you to put them right. The exercises in the text let you practise scientific and language points. Your improved fluency in English will raise your understanding of Earth Science and open up a whole new world of English communication.

Roger Mason

2 February 2018

序

随着中国国际地位的攀升、"一带一路"倡议的推进,国际交流与合作的机会越来越多,中国制造和中国标准亦初见端倪。中国要走向国际,需要越来越多的技术输出,对既懂专业又懂外语的专业技术人才的需求也更为迫切。传统上以通用语言能力为主的大学英语教学活动难以满足国际化专业技术人才的语言需求。因此,不少大学开始改革大学英语教学,增加适应中国国际化战略的专门用途英语(English for Specific Purposes, ESP)教学,针对各专业学科编写的ESP教材也相继出版面世。为配合我校"地学"双一流学科人才培养建设,我们依托我校丰厚的地学英语专业文献,借助地学专家的学术指导,编撰了这套《地学英语系列教材》。

该套系列教材,在阅读版块,分为4册,遵循由简入深的方式,按照ESP教学类别,对所选材料进行分门别类,便于学习者逐级学习使用。教材前两册,即《地学英语阅读教程》和《地学精要》,以一般性学术英语(English for General Academic Purpose, EGAP)为原则编撰。作为学术文献阅读基础,重点介绍地学基础类文章,使学习者对地学类英语文献逐步形成系统的认知,加深对地学类英语文章的文本特色理解,并积累一定的学术文献阅读技巧,为后续两册的学习打下坚实的语言基础,为更高级专业地学学术文献阅读作铺垫。教材后两册,即《地学文献阅读》和《地学语篇语用》,供后续学术英语阅读教学使用,以专用学术英语(English for Specific Academic Purpose, ESAP)为原则编撰。教材摘取真实的学术期刊论文或报告作为阅读文本,使学习者学会快速阅读本专业文献,逐步了解专业文献的写作特点和写作技巧,辅助学生进行专业学术论文写作。通过本系列教材的学习,学习者可以完成由EGAP阅读到ESAP阅读的平稳过渡,既能提高专业文献阅读能力又能为英语论文写作打下一定基础。

本系列阅读版块教材适合作为大学英语的学术英语阅读课程,每册可满足每学期24~36学时的教学需要,也可以作为地学专业双语教学的阅读教材。同时,各册教材重点突出,定位明确,可以作为英语选修课教材,供不同需求层次的学生单独使用。

《地学英语阅读教程》——Basic,基础篇。为EAGP阅读基础版,本册以地学基础类文章为阅读文本,浅显易懂,以阅读了解地学类基础知识为目标。

《地学精要阅读》——Progressive,拓展篇。也是EAGP阅读文本,重点选取与地学相关的专业文本,文体简洁,知识易懂。学习者通过学习本册教程,可以对地学涵盖的相关领域有一个系统的认知。

《地学文献阅读》——Advanced,提高篇。为ESAP阅读文本,重点选取地学专业学术文章或学术报告，用词专业性强，强调专业文本的书写精确性、简洁性，以及文本结构的严谨性。学习者通过本册教程的学习可以对地学专业类的学术文献有更直接的感性认识，通过阅读加深对学术文本的语言特色和写作规范的理解。

《地学语篇语用》——Proficient，专业篇。为 ESAP 加强型阅读文本，选取地学专业期刊论文或学术报告，用词专业性强。本册教程强调学术文献语篇语用的理解，通过语篇建构和语言使用剖析，以及阅读和写作技巧的训练，学习者可以依据论文语篇特色，尝试书写学术研究论文和学术研究报告。

在有条件的情况下，教材可以 4 册连用，也可以根据不同层次学生需求，选取一两册满足教学需要，其他教材作为辅助拓展阅读材料，供学生自学。

由于地学英语学术文本的特殊性，专业词汇难以上口，我们在每册书的后面都提供了加注音标的词汇总表，便于学习者查阅朗读。本系列教材同时提供立体化教材资源服务，既有纸质版教材书本便于携带阅读，也有 MOOC 作为教学辅导，还可以通过扫描二维码链接语音资源。

此系列教材只是我们进行大学英语 ESP 教学改革的第一步，还会面临各种挑战。我们相信，只要我们加倍努力，不断提高学术英语和地学专业知识素养，不断更新改进我们的学术英语阅读教材，一定会为专门用途英语的教学打开一片天地。

2017 年 12 月

前　言

专门用途英语(English for Specific Purposes，ESP)的教学成为我国高校，特别是理工科院校英语改革及发展的方向。《地学英语阅读教程》秉承ESP教学理念进行编写。本书编写的目的在于丰富地质英语的ESP教材，提高地质、资源、环境、工程等相关专业学生国际化的迫切需求，及提高相关专业学生和从业人员的阅读能力。

《地学英语阅读教程》共8个单元，每个单元包括4个部分：单元导读(Preview)、A篇课文精读和练习、阅读技巧介绍(Reading Skills)、B篇雅思题型阅读与练习。其中，A篇阅读材料部分节选自英文原版文章，部分节选自维基百科；B篇阅读材料和练习选自历年雅思考试真题或者模拟题，是A篇阅读材料的进阶版。为了便于教学和自学复习，每篇阅读材料都可以通过扫描二维码的方式获取相应课文和词汇的音频，所有音频的录制均由曾在中国地质大学(武汉)交流的英国学生Katherine Briggs(第1、3、5单元)和Rob Dolan(第2、4、6单元)完成，音频语速保持在120个单词/分钟(正常说话语速，大学英语六级考试听力的语速)左右，保证发音的原汁原味的同时，解决部分复杂专业词汇发音难等问题；书后还附有词汇总表(Glossary)和习题答案(Keys)。本书按照两条主线编写：

(1)每个单元都是按照"普通地质学"这门课程的知识框架来选材，由于我们采取了以下措施，这些地学基础文章变得很友善，难度和大学英语课文相当。

★长度适中，每篇控制在1000字左右；

★生词量除了术语外控制在1.5%内，以减轻学生阅读科普文章的压力(客观地说，除了地学类术语外，词汇量很小)；

★课文虽然是地学科普文章，但是没有专业背景知识的读者也能理解。

(2)阅读方法介绍、雅思题型阅读与练习是本书的另一大特色，本书很好地将地学基础阅读材料、阅读方法与雅思训练相结合。

本书是地质、资源、环境、工程等相关专业的学生们从通用大学英语过渡到专业学术英语学习的"专门用途英语"，为1个学期24～36学时的教学使用。本书可以用于以上相关专业本科生和研究生大学英语"阅读进阶"的主要教材，很好地帮助学生从通用英语学习过渡到专业英语学习的同时，为专业学术论文英文摘要等的写作打下基础。本书也是以上相关专业的专业英语基础阅读内容，可以很好地作为《普通地质学》的辅助英文教材和"普通地质学"双语教学的补充教材。本书亦可作为雅思阅读训练的教材。本书特别为以下读者而编写：

★地质、资源、环境、工程等相关专业的本科生和研究生,特别适合本科生第3、4学期和研究生第1、2学期的英语阅读;

★地质、资源、环境、工程等相关从业人员,作为相关从业人员的专业英语基础阅读材料;

★想提高英语阅读技能、了解地学基础知识的学生;

★雅思备考人员。

本书由赖小春统稿,各单元的编写分工如下。

前言:由赖小春编写,葛亚非、任利民提出修改意见;

第1单元　地球科学简介:由汪卫红编写,并由胡冬梅审阅;

第2单元　地球历史与生命进化:由王晓婧编写和审阅;

第3单元　地球板块与地球结构:由汪卫红编写,并由胡冬梅审阅;

第4单元　元素、矿物和岩石:由曾艳霞编写和审阅;

第5单元　地貌学与地理学:由赖小春编写和审阅;

第6单元　大气圈与气候变化:由付蕾编写和审阅;

第7单元　自然资源与环境保护:由付蕾编写,并由刘雪莲审阅;

第8单元　地质灾害与地质工程:由曾艳霞编写和审阅;

词汇总表:在各单元的编写老师分别整理的基础上,由赖小春和付蕾查重;

习题答案:由各单元的编写老师分别整理和审阅。

本书能够顺利出版,首先,十分感谢中国地质大学(武汉)外国语学院、教务处和地球科学学院的大力支持和帮助。同时,葛亚非、任利民、马昌前、喻建新、肖珊等老师为该书的编写多次提供了专业的指导和建议,任利民老师多次为本书的封面设计提供了大量精美的野外实景图片,Katherine Briggs 和 Rob Dolan 为本书提供了专业的课文和词汇录音,在此一并表示诚挚的感谢。

我们付出了辛勤的努力,但由于水平有限,书中难免存在差错或者不当之处,敬请读者批评指正。

编　者

2017 年 12 月

于地大南望山

Contents

Unit 1　Introduction to Earth Science　*1*
　　Passage A　The Science of Geology　*1*
　　Reading Skills　Vocabulary in Context　*10*
　　Passage B　Earth System Science　*12*

Unit 2　Earth History and Life Evolution　*20*
　　Passage A　Dating Rocks and Fossils Using Geologic Methods　*20*
　　Reading Skills　Signpost Language (Reading Road Map)　*28*
　　Passage B　Fossil Files—"The Palaeobiology Database"　*28*

Unit 3　Plate Tectonics and Earth Structure　*34*
　　Passage A　Continental Drift: an Idea Before Its Time　*34*
　　Reading Skills　Signpost Language　*43*
　　Passage B　Plate Tectonics　*46*

Unit 4　Mineral and Rock　*51*
　　Passage A　Mineral　*51*
　　Reading Skills　Examples as Contextual Clues　*59*
　　Passage B　Classification of Rocks　*60*

Unit 5　Geomorphology and Geography　*65*
　　Passage A　Geological Process—Weathering and Erosion　*65*
　　Reading Skills　Recognising Differences Between Facts and Opinions　*73*
　　Passage B　Disappearing Delta　*74*

Unit 6　Climate Change and Atmosphere　*79*
　　Passage A　Effect of Climate Change　*79*
　　Reading Skills　Guessing Meaning from Context　*85*
　　Passage B　Sun's Fickle Heart May Leave Us Cold　*86*

Unit 7 Natural Resources and Environment Protection *90*
    Passage A   Oil and Gas   *90*
    Reading Skills   Skimming and Scanning   *96*
    Passage B   Tidal Power   *96*

Unit 8 Geological Hazard   *100*
    Passage A   Tsunami   *100*
    Reading Skills   Understanding the Sentences with Negative Words   *107*
    Passage B   Earthquake   *108*

Glossary   *113*

Keys   *129*

Unit 1
Introduction to Earth Science

 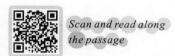

Scan and read along the passage

The spectacular eruption of a volcano, the terror brought by an earthquake, the magnificent scenery of a mountain valley, and the destruction created by a landslide all are subjects for the geologist. The study of geology deals with many fascinating and practical questions about our physical environment. Will there soon be another great earthquake in Sichuan? What was the Ice Age like? Will oil be found if a well is drilled at this location? All these are concerns of geology. In this unit, we will examine the science of geology. We will also briefly trace the development of geology, the major areas of geologic studies, and the current advancement of earth system science.

Passage A

The Science of Geology[①]

Scan and read along the vocabulary

A The word "geology" is from the Greek "geo" (meaning "earth") and "logos" (meaning "discourse"). It is the science that pursues an understanding of planet Earth. Geology is traditionally divided into two broad areas—physical and historical. **Physical geology** examines the materials composing Earth and seeks to understand the many processes that operate beneath and upon its surface. The aim of **historical geology**, on the other hand, is to understand the origin of Earth and its development through time. Thus, it strives to establish a **chronological** arrangement of the **multitude** of physical and biological changes that have occurred in the geologic past. The study of physical geology logically **precedes** the study

physical geology　n. 普通地质学
historical geology　n. 地史学
chronological　adj. 按年代顺序排列的;依时间前后排列记载的
multitude　n. 群众;多数
precede　vt. 领先,在……之前;优于,高于;
vi. 领先,在前面

[①] This passage is adapted from: Lutgens F K, Tarbuck E J. An introduction to geology [M]//Essentials of geology. 11 th ed. Upper Saddle River, NJ, USA: Pearson, 2012: 1 - 35.

of Earth history because we must first understand how Earth works before we attempt to **unravel** its past. It should also be pointed out that physical and historical geology are divided into many areas of specialisation. Table 1 provides a partial list.

Table 1　Different areas of geologic study[1]

Archaeological Geology	History of Geology	Petrology
Biogeosciences	Hydrogeology	Planetary Geology
Engineering Geology	Medical Geology	Sedimentary Geology
Forensic Geology	Mineralogy	Seismology
Geochemistry	Ocean Sciences	Structural Geology
Geomorphology	Paleoclimatology	Tectonics
Geophysics	Paleontology	Volcanology

Note: Many of these areas of study represent interest sections and specialties of associated societies affiliated with the Geological Society of America and the American Geophysical Union, two professional societies to which many geologists belong.

Geology, people, and the environment

B　Many of the problems and issues addressed by geology are of practical value to people. Natural **hazards** are a part of living on Earth. Every day they **adversely** affect millions of people worldwide and are responsible for **staggering** damages. Among the hazardous earth processes studied by geologists are volcanoes, floods, Earthquakes, and **landslides**. Of course, geologic hazards are simply natural processes. They become hazards only when people try to live where these processes occur. According to the United Nations, in 2008, for the first time, more people lived in cities than in rural areas. This global trend toward **urbanisation** concentrates millions of people into **megacities**, many of which are **vulnerable** to natural hazards. Coastal sites are becoming more vulnerable because development often destroys natural defenses such as wetlands and **sand dunes**. In addition, there is a growing threat associated with human influences on earth system such as sea level rise that is linked to global climate change. Other megacities are exposed to **seismic** and volcanic hazards where inappropriate land

unravel　*vt.* 解开；阐明；解决；拆散
vi. 解决；散开
hazard　*n.* 危险；灾害
adversely　*adv.* 不利地；逆地；反对地
straggering　*adj.* 难以置信的，令人震惊的
landslide　*n.* 山崩；滑坡
urbanisation　*n.* 都市化；城市化
megacity　*n.* （人口超过千万的）大城市；巨型城市；特大都市
vulnerable　*adj.* 易受攻击的，易受伤害的；有弱点的；脆弱的
sand dunes　*n.* 沙丘
seismic　*adj.* 地震的；因地震而引起的

use and poor construction practices, coupled with rapid population growth, are increasing vulnerability.

C Resources represent another important focus of geology that is of great practical value to people. They include water and soil, a great variety of **metallic** and nonmetallic minerals, and energy. Together they form the very foundation of modern civilisation. Geology deals not only with the formation and occurrence of these vital resources, but also with maintaining supplies and the environmental impact of their **extraction** and use.

D Not only do geologic processes have an impact on people, but we humans can dramatically influence geologic processes as well. For example, river flooding is natural, but the **magnitude** and frequency of flooding can be changed significantly by human activities such as clearing forests, building cities, and constructing dams. Thus, the impact of human activities on geologic processes is also an important area of geologic studies.

Historical notes about geology

E The nature of our earth—its materials and its processes—has been a focus of study for centuries. Writings about **fossils**, **gems**, earthquakes, and volcanoes date back to the Greeks, more than 2300 years ago. Certainly, the most influential Greek philosopher was Aristotle. Unfortunately, Aristotle's explanations about the natural world were not derived from keen observations and experiments, as is modern science. Instead, they were **arbitrary** pronouncements based on the limited knowledge of his day. He believed that rocks were created under the "influence" of the stars and that earthquakes occurred when air in the ground was heated by central fires and escaped explosively! When confronted with a fossil fish, he explained that "a great many fishes live in Earth motionless and are found when **excavations** are made". Though Aristotle's explanations were inadequate, they continued to be expounded for many centuries, thus **thwarting** the acceptance of more up-to-date ideas.

metallic *adj.* 金属的,含金属的
extraction *n.* 提取,抽出;精炼
magnitude *n.* 大小;量级;〈地震〉震级
fossil *n.* 化石
gem *n.* 宝石
arbitrary *adj.* 任意的;武断的
excavation *n.* 挖掘,发掘
thwart *vt.* 挫败;反对;阻碍

Catastrophism

F In the seventeenth and eighteenth centuries, the **doctrine** of **catastrophism** strongly influenced people's thinking about Earth. Briefly stated, catastrophists believed that Earth's **landscapes** had been shaped primarily by great catastrophes. Features such as mountains and **canyons**, which today we know take great periods of time to form, were explained as having been produced by sudden and often worldwide disasters caused by unknown forces that were no longer in operation.

The birth of modern geology

G In 1795, a Scottish physician and gentleman farmer named James Hutton published *Theory of the Earth*. In this work, Hutton put forth a fundamental principle that is a pillar of modern geology: **uniformitarianism**[2]. It suggests that the physical, chemical, and biological laws that operate today also operated in the geologic past. In other words, the forces and processes that we observe shaping our planet today have been at work for a very long time. Thus, to understand ancient rocks, we must first understand present-day processes and their results. The idea is commonly stated as "The present is the key to the past". Prior to Hutton, no one had effectively demonstrated that geological processes can continue over extremely long periods of time. Hutton persuasively argued that forces that appear small could, over long spans of time, produce effects just as great as those resulting from sudden catastrophic events. Hutton carefully cited verifiable observations to support his ideas. For example, when he argued that mountains are **sculpted** and ultimately destroyed by weathering and the work of running water, and that their **wastes** are carried to the oceans by processes that can be observed, Hutton stated, "we have a chain of facts which clearly demonstrates … that the materials of the wasted mountains have traveled through the rivers", and "there is not one step in all this process … that is not to be actually perceived".

Geology today

H Today, the basic **tenets** of uniformitarianism are just as viable as in Hutton's day. We realise more strongly than ever that the

---

doctrine *n.* 主义;学说;教义;信条
catastrophism *n.* 灾变说;劫数难逃论
landscape *n.* 风景,山水画;地形
canyon *n.* 峡谷
uniformitarianism *n.* 均变论;推今及古原理
sculpt *v.* 造型,雕刻;使成形
waste *n.* 浪费;废物;荒地;损耗;地面风化物
tenet *n.* 原则;信条;教义

present gives us insight into the past and that the physical, chemical, and biological laws that govern geologic processes remain unchanged through time. However, we also understand that the doctrine should not be taken into literally. To say that geological processes in the past were the same as those occurring today is not to suggest that they always had the same relative importance or operated at precisely the same rate. Moreover, some important geologic processes are not currently observable, but evidence that they occur is well established. For example, we know that Earth has experienced impacts from large **meteorites** even though we have no human witnesses. Such events altered Earth's crust, modified its climate, and strongly influenced life on the planet.

❶ The acceptance of uniformitarianism meant the acceptance of a very long history for Earth. Although processes vary in their intensity, they still take a very long time to create or destroy major landscape features. For example, geologists have established that mountains once existed in portions of present-day Minnesota, Wisconsin, Michigan, and Manitoba[3]. Today the region consists of low hills and plains. **Erosion** gradually destroyed these peaks. The rock record contains evidence that shows Earth has experienced many such cycles of mountain building and erosion. It is important to remember that although many features of our physical landscape may seem to be unchanging in terms of the decades over which we observe them, they are nevertheless changing, but on time scales of hundreds, thousands, or even many millions of years.

meteorite　n. 陨星；流星
erosion　n. 侵蚀，腐蚀

Notes

(1) Different areas of geologic study：表 1 中列出的很多研究领域反映了美国地质协会（Geological Society of America, www.geosociety.org）和美国地球物理联合会（American Geophysical Union, www.agu.org）这两大专业地质学家组织下各科研机构的研究领域和侧重点。表 1 中各研究领域的中英文对应分别为：

Archaeological Geology 考古地质	History of Geology 地质学史	Petrology 岩石学
Biogeosciences 生物地球科学	Hydrogeology 水文地质学	Planetary Geology 行星地质学
Engineering Geology 工程地质	Medical Geology 医学地质	Sedimentary Geology 沉积地质学
Forensic Geology 法庭地质学	Mineralogy 矿物学	Seismology 地震学

Geochemistry 地球化学	Ocean Sciences 海洋科学	Structural Geology 构造地质学
Geomorphology 地貌学	Paleoclimatology 古气候学	Tectonics 构造学
Geophysics 地球物理	Palaeontology 古生物学	Volcanology 火山学

(2) Catastrophism and Uniformitarianism：灾变论与均变论。二者均为地质学理论，前者认为地球曾经遭受许多短暂的灾难，其中有些是世界性的。这一思想可以追溯到圣经中的大洪水故事，直至 19 世纪早期，仍有学者主张灾变论，但后来逐渐被另一派认为地球历史长远且渐进的均变论（或称渐变论）所取代。均变论由英国人杰姆斯·赫顿（James Hutton）在 1785 年和 1789 年提出的渐变论（Actualism）衍生而来。均变论基本思想与渐变论大致相同，同时提出了地质时间的概念，并否定了达尔文的天择说。均变论核心要义即"现在是通往过去的一把钥匙"，表示一切过去所发生的地质作用都和现在正在进行的作用方式相同，所以研究现在正在进行的地质作用，就可以明了过去的地球历史。

(3) Minnesota, Wisconsin, Michigan, and Manitoba：北美洲地名，分别为美国中北部的明尼苏达州、中西部的威斯康辛州、北部的密歇根州，以及加拿大中南部的马尼托巴湖。

Online Resources

1. Geological Society of America（GSA，美国地质协会）. It is a global professional society with a growing membership of more than 26 000 individuals in 115 countries. GSA provides access to elements that are essential to the professional growth of earth scientists at all levels of expertise and from all sectors: academic, government, business, and industry. The society unites thousands of earth scientists from every corner of the globe in a common purpose to study the mysteries of our planet (and beyond) and share scientific findings. For more information, please visit http://www.geosociety.org/GSA/.

2. American Geophysical Union（AGU，美国地球物理联合会）. It is a nonprofit organisation of geophysicists, consisting of over 62 000 members from 144 countries. AGU's activities are focused on the organisation and dissemination of scientific information in the interdisciplinary and international field of geophysics. The geophysical sciences involve four fundamental areas: atmospheric and ocean sciences; solid-Earth sciences; hydrologic sciences; and space sciences. The organisation's headquarter is located on Florida Avenue in Washington, D. C. For more information, please visit http://sites.agu.org/.

3. James Hutton（杰姆斯·赫顿）. James Hutton was born on June 3, 1726 in Edinburgh, Scotland, and died on March 26, 1797. Even though James Hutton did not have a degree in Geology, his experiences on his farm gave him the focus to really study the subject and come up with theories about the formation of Earth that were novel at the time. Hutton hypothesised that the interior of Earth was very hot and the processes that changed Earth long ago were the same processes that were at work on Earth in present day. He published his ideas in the book *Theory of the Earth* in 1795. In this book, Hutton even went on to assert that life also followed this pattern. The ideas put forth in the book about life changing over time using the same mechanisms since the beginning of time was in line with the idea of evolution long before Charles Darwin came up with the theory of *Natural Selection*. For more information, please visit https://www.thoughtco.com/about-james-hutton-12.

Exercises

Detailed Understanding

I. Answer the following questions according to the passage you have read.

1. What is geology? Please give a definition of it.
2. What are the two major subdivisions of geology?
3. How does physical geology differ from historical geology?
4. Can you list three different geologic hazards mentioned in the passage?
5. What is the practical value of geologic studies to people?
6. What is Aristotle's theory of geology? Can you illustrate with an example? What influence does his theory has on the development of geology?
7. What are catastrophism and uniformitarianism talking about respectively?
8. Contrast catastrophism and uniformitarianism. What is the fundamental difference between them? How does each view the age of Earth?

Vocabulary

II. Fill in the blanks with the words given below. Change the form where necessary. (15 个单词, 10 个空)

compose	scale	arbitrary	literally	unravel
hazard	tenet	vulnerable	erosion	impact
adversely	construction	landslide	choronological	confront

1. Earthquakes often cause dramatic changes at Earth's surface. In addition to the ground movements, other surface effects include changes in the flow of groundwater, _____, and mudflows.
2. The lack of rainfall causes soils to dry out, making them lighter and more _____ to wind erosion.
3. Natural processes such as earthquakes, landslides, floods, volcanic eruptions, tornadoes, and hurricanes are considered _____ when they occur in populated areas.

4. Obviously, an unarmed person would be in great danger if _____ in the open by an 800-pound bear. However, taking a long-term view, polar bears are much more at risk from climate change than humans are.
5. The construction of dams and reservoirs may result in _____ on the ecosystem.
6. A volcano is considered active if it has erupted within recent recorded history. If a volcano has little _____ and looks fairly fresh, it is considered dormant with the ability to become active again at any time. If a volcano has not erupted within recorded time and is to a large extent, it is thought to be extinct and very unlikely to erupt again.
7. Over the past two centuries, humankind has unintentionally embarked on a global experiment in climate modification that may _____ affect natural systems all over the planet.
8. Igneous rocks are of a variety of minerals, each _____ with unique chemical properties.
9. The _____ sequence of terrane accretion (地体增生) onto a continent can be determined from geologic events that postdate accretion and link adjacent terranes.
10. The VDAP team members are collecting data and analyzing the geology surrounding the volcano to _____ the history of eruptions in the region.

Sentence Structure

Ⅲ. Combine the following sentences with proper connectives to indicate the connections between them.

> **Model**
> Throughout its long existence, Earth has been changing.
> It is changing as you read this passage and will continue to do so into the foreseeable future.
> ⟶ Throughout its long existence, Earth has been changing. In fact, it is changing as you read this passage and will continue to do so into the foreseeable future.

1. The particles in liquids vibrate vigorously.
   Some particles can gain sufficient energy to escape the liquid.

2. Rocks on level areas are likely to remain in place over time.
   The same rocks on slopes tend to move as a result of gravity.

3. If a well is not drilled deep enough, it may stop producing water in periods of prolonged drought.
   Money is wasted if the well is drilled beyond where the well could conceivably ever go dry.

4. All of the planet's forests account for approximately three-quarters of all the biomass on one-fifth of the land.
   Deserts cover about the same land area but account for less than two percent of the biomass.

5. Hurricanes can form under appropriate conditions all over the world, but are named differently in different oceans.
   They are called typhoons in the Pacific Ocean and cyclones in the Indian Ocean.

Translation

IV. Translate the following sentences into Chinese.

1. The aim of historical geology, on the other hand, is to understand the origin of Earth and its development through time. Thus, it strives to establish a chronological arrangement of the multitude of physical and biological changes that have occurred in the geologic past.

2. The study of physical geology logically precedes the study of Earth history because we must first understand how Earth works before we attempt to unravel its past.

3. Other megacities are exposed to seismic and volcanic hazards where inappropriate land use and poor construction practices, coupled with rapid population growth, are increasing vulnerability.

4. Features such as mountains and canyons, which today we know take great periods of time to form, were explained as having been produced by sudden and often worldwide disasters caused by unknown forces that were no longer in operation.

5. It is important to remember that although many features of our physical landscape may seem to be unchanging in terms of the decades over which we observe them, they are nevertheless changing, but on time scales of hundreds, thousands, or even many millions of years.

V. Translate the following passage into English.

查尔斯·莱尔爵士(Sir Charles Lyell)、杰姆斯·赫顿(James Hutton)、阿尔弗雷德·魏格纳(Alfred Wegener),还有哈雷·赫斯(Harry Hess)都是历史上著名的地质学家。他们共同努力促成了板块构造说(Plate Tectonics)、海地扩张说等思想、理论、研究的形成,以及灾变论向均变论的发展。其中,杰姆斯·赫顿于1726年6月3日生于苏格兰爱丁堡,卒于1797年3月26日,被誉为"现代地质学之父",均变论的首创者,其思想直接影响后来的查尔斯·莱尔爵士和查尔斯·达尔文。

Reading Skills

Vocabulary in Context[①]

Vocabulary in context refers to the sentences or the whole paragraph surrounding an unfamiliar word. Context clues can be used to make a good guess at the word's meaning. Here list six different types of context clues:

[①] This section is adapted from webpage from: Independence High School English Department. Vocabulary in context[EB/OL]. [2017-06-12]. http://staff.esuhsd.org/danielle/english%20department%20lvillage/CAHSEE%20English/VocabinContext/VocabinContext.html.

- definition/restatement
- example
- synonym/antonym
- comparison
- contrast
- cause and effect

In the following, you will read and practise using each type. The practices will sharpen your skills in recognising and using context clues. They will also help you add new words to your vocabulary.

1. Definition/Restatement

   Sometimes, writers restate a word in order to define it.

   e.g., When Henry Gonzalez was elected to Congress, many of his Spanish speaking *constituents*, the voters in his district, felt he would fight for their rights.

   The phrase follows "*constituents*" works as a definition to it.

2. Example

   Examples may suggest the meaning of an unknown word.

   e.g., The *adverse* effects of this drug, including dizziness, nausea, and headaches, have caused it to be withdrawn from the market.

   The examples—dizziness, nausea, and headaches—helped you figure out that the word "*adverse*" means "harmful".

3. Synonym/Antonym

   Look for familiar words that may be synonyms or antonyms of words unknown.

   A synonym is a word that means the same or almost the same as the unknown word. In contrast, an antonym is a word that means the opposite of another word. Both may appear elsewhere in a passage to provide meaning either similar or opposite to the unknown one.

   e.g., My doctor said smoking could *terminate* my life. But I told him, "Everyone's life has to end some time."

   The word "end" is a synonym of "*terminate*".

   e.g., I prefer the occasional disturbance of ear-splitting thunder to the *incessant* dripping of our kitchen sink.

   The word "occasional" is an antonym of "*incessant*".

4. Comparison/Contrast

   An unknown word may be compared to or contrasted with a more common word.

   e.g., As in so many polluted cities, the air in our community is sometimes too *contaminated* to breathe.

Here "*contaminated*" is compared to "polluted" mentioned earlier in the sentence.

e.g., The team's uniforms were *immaculate* before the game, but by the end of the first half they were filthy.

Here "*immaculate*" is contrasted with a more familiar word "filthy".

5. Cause and Effect

An unfamiliar word may be related to the cause or effect of an action, feeling, or idea.

e.g., Will Rogers was considered to be a *humanitarian* because he worked to improve people's lives.

From the reasons given, we can infer that a "*humanitarian*" is a person who "works to improve people's lives".

Textbook authors typically set important words in italic or boldface and define those words for you, often providing examples as well. Sometimes, they may even explain them through comparison and contrast, or cause and effect to make the meanings explicit to readers. While reading, we should try our best to figure out the meanings of unfamiliar words by looking at their context—the words surrounding them.

Passage B

Scan and read along the passage

Earth System Science[①]

A Earth System Science is the science that studies the whole planet as a system of **innumerable** interacting parts and focuses on the changes within and among those parts. Examples of these parts are oceans, the atmosphere, continents, lakes and rivers, soils, plants, and animals; each can be studied separately, but each is dependent on and **interconnected** with the others. The global interconnectedness of air, water, rocks, and life has become a focus of modern scientific investigation. As a result, a new approach to the study of Earth has taken hold. The traditional way to study Earth has been to focus on separate units—a population of animals, the atmosphere, a lake, a single **mountain range**, soil in some region—in isolation from other units. In the new **holistic**

Scan and read along the vocabulary

innumerable *adj.* 无数的,不计其数的,数不清的
interconnected *adj.* 连通的;有联系的
mountain range *n.* 山脉;山岳地带,山地
holistic *adj.* 整体的;全盘的

[①] This Passage is adapted from: Skinner B J, Murck B. Earth system [M]//The blue planet: An introduction to earth system science. 3rd ed. Hoboken, NJ, USA: Wiley, 2011: 5 – 28.

approach, Earth is studied as a whole and is viewed as a system of many separate but interacting parts. Nothing on Earth is isolated; research reveals **numerous interactions** among all of the parts.

Ⓑ The scope of Earth Science is vast. This broad field can be broken into five major areas of specialisation: astronomy, meteorology, geology, oceanography, and environmental science. The study of objects beyond Earth's atmosphere is called **astronomy**. Prior to the invention of sophisticated instruments, such as telescope, many astronomers merely described the locations of objects in space in relation to each other. Today, Earth scientists study the universe and everything in it, including **galaxies**, stars, planets, and other bodies they have identified. The study of the forces and processes that cause the atmosphere to change and produce weather is **meteorology**. Meteorologists try to **forecast** the weather and learn how changes in weather over time might affect Earth's climate. The study of the materials that make up Earth, the processes that form and change these materials, and the history of the planet and its life-forms since its origin is the branch of earth science known as geology. Geologists identify rocks, study **glacial** movements, interpret clues to Earth's 4.6-billion-year history, and determine how forces change our planet. The study of Earth's oceans, which cover nearly three-fourths of the planet, is called **oceanography**. Oceanographers study the creatures that **inhabit** salt water, measure different physical and chemical properties of the oceans, and observe various processes in these bodies of water. The study of the interactions of organisms and their surroundings is called environmental science. Environmental scientists study how **organisms** impact the environment both positively and negatively. The topics an environmental scientist might study include natural resources, pollution, alternative energy sources, and the impact of humans on the atmosphere.

Ⓒ The study of our planet is a broad endeavor, and as such, each of the five major areas of Earth Science consists of a variety of subspecialties, some of which are listed in Table 1.

numerous *adj.* 许多的,很多的
interaction *n.* 相互作用;交互作用
astronomy *n.* 天文学
galaxy *n.* 银河;星系;银河系;一群显赫的人
meteorology *n.* 气象状态,气象学
forecast *vt.* 预报,预测;预示;
vi. 进行预报,作预测;
n. 预测,预报;预想
glacial *adj.* 冰的;冰冷的;冰河时代的
oceanography *n.* 海洋学
inhabit *vt.* 栖息;居住于;占据;
vi. 居住;栖息
organism *n.* 有机体;生物体;微生物

Table 1 Subspecialties of Earth Science

Major Area of Study	Subspecialty	Subjects Studied
Astronomy	**Astrophysics**	Physics of the universe, including the physical properties of objects found in space
Astronomy	**Planetary Science**	Planets of the solar system and the processes that form them
Meteorology	**Climatology**	Patterns of weather over a long period of time
Meteorology	**Atmospheric Chemistry**	Chemistry of Earth's atmosphere, and the atmospheres of other planets
Geology	**Paleontology**	Remains of organisms that once lived on Earth; ancient environments
Geology	**Geochemistry**	Earth's composition and the processes that change it
Oceanography	**Physical Oceanography**	Physical characteristics of oceans, such as waves, and currents
Oceanography	**Marine Geology**	Geologic features of the ocean floor, including plate tectonics of the ocean
Environmental Science	Environmental Soil Science	Interactions between human and the soil, such as the impact of farming practices; effects of pollution on soil, plants, and groundwater
Environmental Science	Environmental Chemistry	Chemical alterations to the environment through pollution and natural means

Note: For more information about the scope of Earth Science, visit http://www.glencoe.com.

astrophysics *n.* 天体物理学
planetary science *n.* 行星科学
climatology *n.* 气候学;风土学
atmospheric chemistry *n.* 大气化学
palaeontology *n.* 古生物学
geochemistry *n.* 地球化学
physical oceanography *n.* 海洋物理学
marine geology *n.* 海洋地质学（亦作 geological oceanography）

D Scientists who study Earth have identified four main Earth systems: the **geosphere**, the **atmosphere**, the **hydrosphere**, and the **biosphere**. Each system is unique, yet each interacts with the others.

E The area from the surface of Earth down to its centre is called the geosphere. The geosphere is divided into three main parts: **crust**, **mantle**, and **core**. As illustrated in Figure 1, the rigid outer shell of Earth is called the crust. There are two kinds of crust—continental crust and oceanic crust. The oceanic crust is roughly 7km (5 miles) thick and composed of the dark **igneous** rock basalt. By contrast, the continental crust averages about 35km (22 miles) thick but may exceed 70km (40 miles) in some mountainous regions such as **the Rockies** and **the Himalayas**. Unlike the oceanic crust, which has a relatively **homogeneous** chemical **composition**, the continental crust consists of many rock types. Although the upper crust has an average composition of **granodiorite**, it varies considerably from place to place. Just below the crust is Earth's mantle. The mantle differs from the crust both in volume and composition. The solid, rocky shell extends nearly 2900km (1800 miles) in depth and contains more than 82% of Earth's volume. The temperature of the mantle is also much warmer than the temperatures found in Earth's crust, ranging from 100°C to 4000°C. Below the mantle is Earth's core. The composition of the core is thought to be an **iron-nickel alloy** with minor amounts of oxygen, **silicon**, and **sulfur**—elements that readily form compounds with iron. The core is divided into two regions that exhibit very different mechanical strengths. The outer core is a liquid layer 2270km (1410 miles) thick. It is the movement of metallic iron within this zone that generates Earth's **magnetic** field. The inner core is a sphere having a **radius** of 1216km (754 miles). Despite its higher temperature, the iron in the inner core is solid due to the immense pressures that exist in the centre of the planet.

F The blanket of gases that surrounds our planet is called the atmosphere. Compared to the thickness (radius) of the solid Earth (about 6400km or 4000 miles), the atmosphere is a very shallow layer. One-half lies below an **altitude** of 5.6km (3.5 miles), and 90% occurs within just 16km (10 miles) of Earth's surface. Despite its modest dimensions, this thin blanket of air is an integral part of the planet. It not only provides the air that we

---

geosphere n. 岩石圈；陆界
atmosphere n. 气氛，大气；空气；大气圈；大气层
hydrosphere n. 水界，水圈；水气
biosphere n. 生物圈
crust n. 外壳；面包皮；坚硬外皮 vi. 结硬皮；结成外壳； vt. 盖以硬皮；在……上结硬皮
mantle n. 地幔；斗篷；覆盖物
core n. 地核
igneous adj. 火的；火成的；似火的；igneous rock 火成岩
the Rockies n. 落基山脉，纵贯加拿大、美国的山脉
the Himalayas n. 喜马拉雅山脉
homogeneous adj. 均质的；同种的；齐次的；相同特征的
composition n. 作文，作曲；构成；合成物；组成
granodiorite n. 花岗闪长岩
iron-nickel alloy n. 铁镍合金
silicon n. 硅；硅元素
sulfur n. 硫磺；硫磺色
magnetic adj. 地磁的；有磁性的；有吸引力的
radius n. 半径，半径范围；界限
altitude n.（地表面、海平面之上的）高，高度；（尤指）海拔，深（度）；垂直距离；平地纬度

breathe, but also acts to protect us from Sun's intense heat and dangerous **ultraviolet radiation**. The energy exchanges that continually occur between the atmosphere and the surface and between the atmosphere and space produce the effects we call weather and climate. If, like Moon, Earth had no atmosphere, our planet would not only be lifeless but many of the processes and interactions that make the surface such a dynamic place could not operate. Without weathering and erosion, the face of our planet might more closely resemble the **lunar** surface, which has not changed **appreciably** in nearly 3 billion years.

Figure 1　Earth's geosphere composition

(The geosphere is composed of everything from the crust to the centre of Earth. Notice how thin the crust is in relation to the rest of the geosphere's components.)

G　All the water on Earth, including the water in the atmosphere, makes up the hydrosphere. About 97% of Earth's water exists as **salt water**, while the remaining 3% is **freshwater** contained in **glaciers**, lakes and rivers, and beneath Earth's surface as **groundwater**. Only a **fraction** of Earth's total amount of freshwater is in lakes and rivers. The hydrosphere is a dynamic mass of water that is continually on the move, **evaporating** from the oceans to the atmosphere, **precipitating** back to the land, and running back to the ocean again. The global ocean is certainly the most **prominent** feature of the hydrosphere, blanketing nearly 71% of Earth's surface to an average depth of about 3800m. Water more than anything else makes Earth unique and brings Earth the name of "the Blue Planet" or "the Blue Marble".

H　The biosphere includes all organisms on Earth as well as the environments in which they live. Most organisms live within a few metres of Earth's surface, but some exist deep beneath the ocean's surface, and others live high **atop** Earth's mountains. All of Earth's life-forms require interaction with at least one of the other systems for their **survival**.

❶ In fact, Earth's biosphere, geosphere, hydrosphere, and atmosphere are all **interdependent** systems (Figure 2). Each is related in some way to the others to produce a complex and continuously interacting whole that we call Earth system. For example, Earth's present atmosphere formed millions of years ago through interactions with the geosphere, hydrosphere, and biosphere. Organisms in the biosphere, including humans, continue to change the atmosphere through their activities and natural processes. Rather than looking through the limited **lens** of only one of the traditional sciences—geology, atmosphere science, chemistry, biology, and so forth—Earth system attempts to **integrate** the know-ledge of several academic fields. By using **interdisciplinary** approach, we hope to achieve the level of understanding necessary to comprehend and solve many of our global environmental problems.

atop　　*prep.* 在……的顶上；
adv. 在顶上
survival　　*n.* 幸存, 残存；幸存者, 残存物
interdependent　　*adj.* 相互依赖的；互助的
lens　　*n.* 镜头；透镜；晶状体
integrate　　*vt.* 使……完整；使……成整体；求……的积分；
vi. 求积分；取消隔离；成为一体；
adj. 整合的；完全的；
n. 一体化；集成体
interdisciplinary　　*adj.* 各学科间的

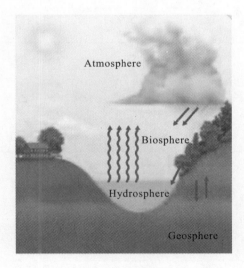

Figure 2　Interdependent Earth systems
(Notice how water from the hydrosphere enters the atmosphere, falls on the biosphere, and soaks into the geosphere.)

Questions 1-4

Passage B has nine paragraphs, A-I.

Choose the correct heading for Paragraphs B-D and H from the list of headings below.

Write the correct number ⅰ－ⅹⅲ in boxes 1－4.

| 1 | | 2 | | 3 | | 4 | |

List of Headings

ⅰ. The atmosphere
ⅱ. The origin of Earth System Science
ⅲ. The astronomy, meteorology, geology, oceanography, and environmental science
ⅳ. The interconnectedness of four spheres
ⅴ. The hydrosphere
ⅵ. The formation of weather and climate
ⅶ. Earth systems
ⅷ. The subspecialties of Earth studies
ⅸ. The internal structure of Earth
ⅹ. The geosphere
ⅺ. The difference between Earth science and geology
ⅻ. The scope of Earth Science
ⅹⅲ. The biosphere

Example	Paragraph **A**	Answer	ⅱ
1	Paragraph **B**	Answer	
Example	Paragraph **C**	Answer	ⅷ
Example	Paragraph **D**	Answer	ⅶ
2	Paragraph **E**	Answer	
3	Paragraph **F**	Answer	
Example	Paragraph **G**	Answer	ⅴ
Example	Paragraph **H**	Answer	ⅹⅲ
4	Paragraph **I**	Answer	

Questions 5－10

Do the following statements reflect the claims of the writer in Passage B?

In boxes 5－10, write

    Y(YES)   if the statement reflects the claims of the writer
    N(NO)   if the statement contradicts the claims of the writer
    NG(NOT GIVEN)   if it is impossible to say what the writer thinks about this

| 5 | | 6 | | 7 | | 8 | | 9 | | 10 | |

5. Paleontology studies ancient environments and the remains of organisms that once lived on Earth.
6. Earth's atmosphere contains about 78% nitrogen and 21% oxygen. The remaining 1% of gases in the atmosphere include water vapor, argon, carbon dioxide, and other trace gases.
7. On average, the oceanic crust is thicker than the continental crust.
8. Earth's atmosphere provides oxygen for living things, protects Earth's inhabitants from harmful radiation of Sun, and helps to keep the planet at a temperature suitable for life.
9. The hydrosphere includes salt water in oceans, freshwater in lakes, rivers and streams, and frozen water found in continental ice sheets and glaciers, but not underground water beneath Earth's surface.
10. The biosphere includes the hydrosphere, crust, and the atmosphere. It is located above the deeper layers of Earth.

Questions 11—13

Complete the summary of Paragraphs B—G with the list of words A—H below.

Write the correct letter A—H in boxes 11—13.

| 11 | | 12 | | 13 | |

A. outer	B. gaseous	C. external	D. spheres
E. realms	F. combine	G. sections	H. integrate

Earth's physical environment is traditionally divided into three major parts: the solid Earth, the geosphere; the water portion of our planet, the hydrosphere; and Earth's __11__ envelope, the atmosphere. In addition, the biosphere, the totality of life on Earth, interacts with each of the three physical realms and is an equally integral part of Earth. Although each of Earth's four __12__ can be studied separately, they are all related in a complex and continuously interacting whole that we call the "Earth system". Earth system science uses an interdisciplinary approach to __13__ the knowledge of several academic fields in the study of our planet and its global environmental problems.

Unit 2
Earth History and Life Evolution

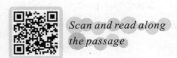

Scan and read along the passage

The geological history of Earth follows the major events in Earth's past based on the geologic time scale, a system of chronological measurement based on the study of the planet's rock layers (stratigraphy). Fossils are the preserved remains or traces of animals, plants, and other organisms from the remote past. The totality of fossils, both discovered and undiscovered, and their placement in fossiliferous (fossil-containing) rock formations and sedimentary layers (strata) is known as the fossil record. The development of dating techniques allowed geologists to determine the numerical or "absolute" age of the various strata and thereby the included fossils.

Passage A

Dating Rocks and Fossils Using Geologic Methods[①]

Scan and read along the vocabulary

Ⓐ There are three general approaches that allow scientists to date geological materials and answer the question: "How old is this fossil?"

Relative dating to determine the age of rocks and fossils

Ⓑ Geologists have established a set of principles that can be applied to **sedimentary** and **volcanic** rocks that are exposed at Earth's surface to determine the relative ages of geological events preserved in the rock record.

Ⓒ This is the principle of original **horizontality**: layers of **strata** are **deposited** horizontally or nearly horizontally. The principle of

sedimentary　*adj.* 沉积的,沉淀性的
volcanic　*adj.* 火山的;猛烈的;暴烈的
n. 火山岩
horizontality　*n.* 水平状态(或位置、性质)
strata　*n.* 地层;岩层(stratum 的名词复数);社会阶层
deposit　*vi.* 沉淀
vt. 储蓄;寄存;放置,安置;付保证金
n. 保证金;储蓄,存款;沉淀物;寄存,寄存品

[①] This passage is adapted from: Peppe D J, Deino A L. Dating rocks and fossils using geologic methods[J]. Nature Education Knowledge, 2013, 4(10): 1.

superposition builds on the principle of original horizontality. The principle of superposition states that in an **undeformed** sequence of sedimentary rocks, each layer of rock is older than the one above it and younger than the one below it. Accordingly, the oldest rocks in a sequence are at the bottom and the youngest rocks are at the top.

D Sometimes sedimentary rocks are disturbed by events, such as fault movements[1] that cut across layers after the rocks were deposited. This is the principle of cross-cutting[2] relationships. The principle states that any geologic features that cut across strata must have formed after the rocks they cut through.

E The principles of original horizontality, superposition, and cross-cutting relationships allow events to be ordered at a single location. However, they do not reveal the relative ages of rocks preserved in two different areas. In this case, fossils can be useful tools for understanding the relative ages of rocks. Each fossil species reflects a unique period of time in Earth's history. The principle of **faunal succession**[3] states that different fossil species always appear and disappear in the same order, and that once a fossil species goes extinct, it disappears and cannot reappear in younger rocks.

Determining the numerical age of rocks and fossils

F Unlike relative dating methods, absolute dating methods provide chronological estimates of the age of certain geological mate-rials associated with fossils, and even direct age measurements of the fossil material itself. To establish the age of a rock or a fossil, researchers use some type of clock to determine the date it was formed. Geologists commonly use **radiometric** dating methods, based on the natural **radioactive decay** of certain elements such as **potassium** and carbon, as reliable clocks to date ancient events. Geologists also use other methods—such as electron spin **resonance** and thermo luminescence, which assess the effects of radioactivity on the accumulation of electrons in **imperfections**, or "traps", in the crystal structure of a mineral—to determine the age of the rocks or fossils.

G Most **isotopes** found on Earth are generally stable and do not change. However, some isotopes, like ^{14}C, have an unstable nucleus and are radioactive. This means that occasionally the un-

superposition　n. 叠加;叠置;叠覆
undeformed　adj. 无形变的
faunal　adj. 动物区系的
succession　n.〈生〉自然演替;一系列,接连;继承人,继承权,继承顺序
radiometric　adj. 辐射度的,放射性测量的
radioactive　adj. 放射性的
decay　n. 腐败、衰退的状态
　　vt. 衰败,衰退,衰落
　　vi. 腐烂,腐朽
potassium　n.〈化〉钾
resonance　n. 共振;共鸣;反响
imperfection　n. 不完美;缺点,瑕疵
isotope　n.〈化〉同位素

stable isotope will change its number of **protons**, **neutrons**, or both. This change is called radioactive decay. For example, unstable ^{14}C transforms to stable nitrogen (^{14}N). The **atomic nucleus** that decays is called the parent isotope. The product of the decay is called the daughter isotope. In the example, ^{14}C is the parent and ^{14}N is the daughter.

H Some minerals in rocks and organic matter (e.g., wood, bones, and shells) can contain radioactive isotopes. The **abundances** of parent and daughter isotopes in a sample can be measured and used to determine their age. This method is known as radiometric dating.

I The rate of decay for many radioactive isotopes has been measured and does not change over time. Thus, each radioactive isotope has been decaying at the same rate since it was formed, ticking along regularly like a clock. For example, when potassium is **incorporated** into$^{(4)}$ a mineral that forms when **lava** cools, there is no **argon** from previous decay (argon, a kind of gas, escapes into the atmosphere while the lava is still molten). When that mineral forms and the rock cools enough that argon can no longer escape, the "radiometric clock" starts. Over time, the radioactive isotope of potassium decays slowly into stable argon, which **accumulates** in the mineral.

J The amount of time that it takes for half of the parent isotope to decay into daughter isotopes is called the half-life of an isotope$^{(5)}$. When the quantities of the parent and daughter isotopes$^{(6)(7)}$ are equal, one half-life has occurred. If the half-life of an isotope is known, the abundance of the parent and daughter isotopes can be measured and the amount of time that has **elapsed** since the "radiometric clock" started can be calculated.

K **Radiation**, which is a byproduct of radioactive decay, causes electrons to **dislodge** from$^{(8)}$ their normal position in atoms and become trapped in imperfections in the crystal structure of the material. Dating methods like thermo luminescence, optical stimulating luminescence and electron spin resonance, measure the accumulation of electrons in these imperfections, or "traps", in the crystal structure of the material. If the amount of radiation to which an object is exposed remains constant, the amount of electrons trapped in the imperfections in the crystal structure of the

---

proton n.〈物〉质子
neutron n.〈物〉中子
atomic adj. 原子的；原子能的，原子武器的；极微的
nucleus n. 中心，核心；（原子）核；起点，开始；〈微〉细胞核
abundance n. 丰度；丰富，充裕；大量，极多；盈余
incorporate vt. 包含；组成公司；使混合；
vi. 合并；包含；吸收；混合
argon n. 氩
lava n. 熔岩；火山岩
accumulate vt. 堆积，积累
vi.（数量）逐渐增加，（质量）渐渐提高
elapse vi. 消逝；时间过去
n.（时间的）消逝
radiation n. 辐射；放射物；辐射状；分散
dislodge vi. 移走，离开原位
vt. 把……逐出，驱逐；把……移动

material will be proportional to the age of the material. These methods are applicable to materials that are up to about 100 000 years old. However, once rocks or fossils become much older than that, all of the "traps" in the crystal structures become full and no more electrons can accumulate, even if they are dislodged.

Using paleomagnetism to date rocks and fossils

L Earth is like a **gigantic magnet.** It has a magnetic north and south pole and its magnetic field is everywhere. Just as the magnetic needle in a compass will point toward magnetic north, small magnetic minerals that occur naturally in rocks point toward magnetic north, **approximately** parallel to Earth's magnetic field. Because of this, magnetic minerals in rocks are excellent recorders of the orientation, or **polarity**, of Earth's magnetic field.

M Through geologic time, the polarity of Earth's magnetic field has switched, causing **reversals** in polarity. Earth's magnetic field is generated by electrical currents that are produced by **convection** in Earth's core. During magnetic reversals, there are probably changes in convection in Earth's core leading to changes in the magnetic field. Earth's magnetic field has reversed many times during its history. When the magnetic north pole is close to the geographic north pole (as it is today), it is called normal polarity. Reversed polarity is when the magnetic "north" is near the geographic south pole. Using radiometric dates and measurements of the ancient magnetic polarity in volcanic and sedimentary rocks (termed **palaeomagnetism**), geologists have been able to determine precisely when magnetic reversals occurred in the past. Combined observations of this type have led to the development of the Geomagnetic Polarity Time Scale (GPTS[9]). The GPTS is divided into periods of normal polarity and reversed polarity.

N Geologists can measure the paleomagnetism of rocks at a site to reveal its record of ancient magnetic reversals. Every reversal looks the same in the rock record, so other lines of evidence are needed to **correlate** the site to[10] the GPTS. Information such as index fossils or radiometric dates can be used to correlate a particular paleomagnetic reversal to a known reversal in the GPTS. Once one reversal has been related to the GPTS, the numerical age of the entire sequence can be determined.

gigantic *adj.* 巨大的,庞大的
magnet *n.* 磁铁,磁石;有吸引力的人或物;〈物〉磁体
polarity *n.* 〈物〉极性;〈生〉反向性;对立;〈数〉配极
reversal *n.* 倒转,颠倒;反复;逆转,反转
convection *n.* 传送,对流;运流
palaeomagnetism *n.* 古地磁;古磁学
correlate *v.* 联系;使互相关联
n. 相关物;相关联的人
adj. 相关的;相应特点的

Notes

(1) fault movements：断层运动。
(2) cross-cutting：穿插结构。
(3) faunal succession：生物演替。
(4) incorporate into：并入；划归；使成为……的一部分。
(5) half-life of an isotope：同位素半衰期。
(6) parent isotope：母同位素；衰变之前的放射性同位素。
(7) daughter isotope：子同位素；衰变之后的放射性同位素。
(8) dislodge from：把……驱逐出；从……里取出。
(9) GPTS：Geomagnetic Polarity Time Scale，地磁极性时间表。
(10) correlate to：联系；使相互联系。

Online Resources

To know more information about this text, please visit https://www.elsevier.com/books/the-geologic.

Exercises

Detailed Understanding

Ⅰ. Answer the following questions according to the passage you have read.
1. What are the three general geologic approaches to date rocks and fossils?
2. What are the principles to determine the relative ages of geological events preserved in the rock record?
3. In what way are absolute dating methods different from relative dating methods?
4. How do radiometric dating methods work?
5. What are the typical ways to assess the effects of radioactivity on the accumulation of electrons in imperfections, or "traps"?
6. On what occasion is measuring the accumulation of electrons in imperfections not applicable?
7. Why magnetic minerals in rocks are excellent recorders of the orientation, or polarity, of Earth's magnetic field?
8. How to determine the numerical age by using paleomagnetism?

Unit 2 Earth History and Life Evolution

Vocabulary

II. Fill in the blanks with the words given below. Change the form where necessary. (15 个单词, 10 个空)

fossil	deposit	chronological	abundance	incorporate into
accumulate	gigantic	reversal	dislodge from	succession
radiation	elapse	correlate to	lava	strata

1. The area has an _____ of wildlife.
2. The thorn of bone may _____ his throat without surgery.
3. The volcano kept throwing up streams of _____.
4. Forty-eight hours have _____ since his arrest.
5. A _____ task of national reconstruction awaits us.
6. You've also learned how easy it is to _____ a Java Web application using Open ID 4 Java.
7. The _____ sequence gives the book an element of structure.
8. After 10 minutes the surplus material is washed away and any remaining _____ examined with ultra violet light.
9. At this distance of time' it is difficult to date the _____.
10. What he had in mind amounted to nothing less than a total _____ of the traditional role of the executive.

Sentence Structure

III. Combine each pair of the following sentences, using a "once" clause.

> **Model**
> One reversal has been related to the GPTS.
> On that occasion the numerical age of the entire sequence can be determined.
> ⟶Once one reversal has been related to the GPTS, the numerical age of the entire sequence can be determined.

1. He has made up his mind to choose the course.
   In that case he will not be swerved from his course.

2. You find a mistake.
   Then you should correct it.

3. The environmental damage is done.
   It takes many years for the ecosystem to recover.

4. Virtue is lost.
   All is lost.

5. You understand this rule.
   Then you will have no further difficulty.

Translation

Ⅳ. Translate the following sentences into Chinese.

1. Geologists have established a set of principles that can be applied to sedimentary and volcanic rocks that are exposed at Earth's surface to determine the relative ages of geological events preserved in the rock record.

2. Geologists commonly use radiometric dating methods, based on the natural radioactive decay of certain elements such as potassium and carbon, as reliable clocks to date ancient events.

3. Geologists also use other methods—such as electron spin resonance and thermo luminescence, which assess the effects of radioactivity on the accumulation of electrons in imperfections, or "traps", in the crystal structure of a mineral—to determine the age of the rocks or fossils.

4. Radiation, which is a byproduct of radioactive decay, causes electrons to dislodge from their normal position in atoms and become trapped in imperfections in the crystal structure of the material.

5. Geologists can measure the paleomagnetism of rocks at a site to reveal its record of ancient magnetic reversals.

Ⅴ. Translate the following passage into English.

研究化石可以了解生物的演化并能帮助确定地层的年代。化石地质年代测定的方法之一,是相对年代测定法,其原理是新的岩层沉积在较老的岩层之上。因此,如果它们没有被断层等扰乱的话,岩层的相对年代可以由它们在层序中的位置确定。此外,还可以根据生物演变来测定。一般说来,地层年代越新,其中所含生物就越进步、越复杂。另一方面,不同时期的地层中含有不同类型的化石及化石组合,而相同时期且在相同地理环境下所形成的地层,不论相距多远都含有相同化石或化石组合。

Reading Skills

Signpost Language (Reading Road Map)

Signpost words/phrases/sentences can tell you when the author is going to make a list, change the topic, add an example or simply explain their point of view in another way. In general, signpost language makes it easier to identify when the subject matter, direction or perspective of the passage may change.

If you understand and read for these signpost word markers, it becomes much easier to follow what is being said, to get the main idea, to grasp the logic chain. An example in the following Passage B:

"**B** But already the project is attracting harsh criticism. Some experts believe it to be seriously flawed. They point out that a database is only as good as the data fed into it, and that even if all the current fossil finds were catalogued, they would provide an incomplete inventory of life because we are far from discovering every fossilised species. They say that researchers should get up from their computers and get back into the dirt to dig up new fossils. Others are more sceptical still, arguing that we can never get the full picture because the fossil record is riddled with holes and biases."

At the very beginning, the author uses "*But*" to show the opposite opinion of the mentioned project in the above paragraph, namely, Paragraph A in Passage B. It is evident to predict that the whole paragraph will talk about the disagreement with detailed information.

Passage B

 Scan and read along the passage

Fossil Files—"The Palaeobiology Database"[①]

 Scan and read along the vocabulary

A Are we now living through the sixth extinction as our own activities destroy **ecosystems** and wipe out diversity? That's the **doomsday scenario** painted by many ecologists, and they may well be right. The trouble is we don't know for sure because we don't have a clear picture of how life changes between extinction events or what has happened in previous **episodes**. We don't even know

ecosystem   *n.* 〈生〉生态系统
doomsday   *n.* 世界末日;遭灾之日
scenario   *n.* 方案;剧情概要;分镜头剧本
episode   *n.* 插曲;一集;片段

---

① This Passage is adapted from: Borrero F, Hess F S, Hsu J, et al. The nature of science[M]//Earth science: geology, the environment, and the universe. New York Student Edition. New York, NY, USA: McGraw-Hill Education, 2008: 1-27.

how many species are alive today, let alone the rate at which they are becoming extinct. A new project aims to fill some of the gaps. The **Palaeobiology Database aspires** to be an online **repository** of information about every fossil ever dug up. It is a huge **undertaking** that has been described as biodiversity's **equivalent** of the Human Genome Project. Its organisers hope that by recording the history of biodiversity they will gain an insight into how environmental changes have shaped life on Earth in the past and how they might do so in the future. The database may even indicate whether life can rebound no matter what we throw at it, or whether a human induced extinction could be without parallel, changing the rules that have applied throughout the rest of the planet's history.

B But already the project is attracting harsh criticism. Some experts believe it to be seriously flawed. They point out that a database is only as good as the data fed into it, and that even if all the current fossil finds were **catalogued**, they would provide an incomplete inventory of life because we are far from discovering every fossilised species. They say that researchers should get up from their computers and get back into the dirt to dig up new fossils. Others are more **sceptical**, still arguing that we can never get the full picture because the fossil record is riddled with holes and **biases**.

C Fans of the Paleobiology Database acknowledge that the fossil record will always be incomplete. But they see value in looking for global patterns that show relative changes in biodiversity. "The fossil record is the best tool we have for understanding how diversity and extinction work in normal times," says John Alroy from the National Centre for Ecological Analysis and Synthesis in Santa Barbara. "Having a background extinction estimate gives us a **benchmark** for understanding the mass extinction that's currently under way. It allows us to say just how bad it is in relative terms."

D To this end, the Paleobiology Database aims to be the most thorough attempt yet to come up with good global diversity **curves**. Every day between 10 and 15 scientists around the world add information about fossil finds to the database. Since it got up and running in 1998, scientists have entered almost 340 000 specimens, ranging from plants to whales to insects to dinosaurs to sea

---

palaeobiology n. 古生物学
database n. 数据库；资料库
aspire v. 渴望；立志；追求
repository n. 仓库；贮藏室；博物馆
undertaking n. 事业；保证；企业
equivalent adj. 相等的，相当的
catalogue n. 目录，一览表
sceptical adj. 怀疑的；怀疑论者的
biase n. 偏差
benchmark n. 基准，参照；标准检查程序
curve n. 弧线，曲线；曲线状物
v. 使弯曲；使成曲线；使成弧形

urchins. Overall totals are updated hourly at www.paleodb.org. Anyone can download data from the public part of the site and play with the numbers to their heart's content. Already, the database has thrown up some surprising results. Looking at the big picture, Alroy and his colleagues believe they have found evidence that biodiversity reached a **plateau** long ago, contrary to the received wisdom that species numbers have increased continuously between extinction events. "The traditional view is that diversity has gone up and up and up," he says. "Our research is showing that diversity limits were approached many tens of millions of years before the dinosaurs **evolved**, much less suffered extinction." This suggests that only a certain number of species can live on Earth at a time, filling a prescribed number of **niches** like spaces in a **multi-storey** car park. Once it's full, no more new species can squeeze in, until extinctions free up new spaces or something rare and **catastrophic** adds a new floor to the car park.

E Alroy has also used the database to reassess the accuracy of species names. His findings suggest that irregularities in classification inflate the overall number of species in the fossil record by between 32% and 44%. Single species often end up with several names, he says, due to misidentification or poor communication between **taxonomists** in different countries. Repetition like this can distort diversity curves. "If you have really bad taxonomy in one short **interval**, it will look like a diversity spike—a big diversification followed by a big extinction—when all that has happened is a change in the quality of names," says Alroy. For example, his statistical analysis indicates that of the 4861 North American fossil mammal species catalogued in the database, between 24% and 31% will eventually prove to be **duplicates**.

F Of course, the fossil record is **undeniably patchy**. Some places and times have left behind more fossil-filled rocks than others. Some have been sampled more **thoroughly**. And certain kinds of creatures—those with hard parts that lived in oceans, for example—are more likely to leave a record behind, while others, like **jellyfish**, will always remain a **mystery**. Alroy has also tried to account for this. He estimates, for example, that only 41% of North American **mammals** that have ever lived are known from fossils, and he suspects that similar proportion of fossils are

urchin　n.顽童;淘气鬼;〈动〉獢;〈动〉海胆
plateau　n.高原;平稳时期;停滞期
evolve　v.发展;进化;设计
niche　n.壁龛;合适的位置(工作等);商机
multi—storey　adj.多层的
catastrophic　n.灾难的;惨重的
taxonomist　n.分类学者
interval　n.间隔;幕间休息;区间
duplicate　v.重复;复制
　adj.复制的;副本的
　n.副本;完全一样的东西
undeniably　adv.不可否认地
patchy　adj.不调和的;拼凑成的;有补丁的
thoroughly　adv.彻底地;完全地
jellyfish　n.水母;海蜇;软弱无用的人;意志薄弱的人
mystery　n.秘密,谜;神秘
mammal　n.哺乳动物

missing from other groups, such as **fungi** and insects.

G Not everyone is **impressed** with such **mathematical wizardry**. Onathan Adrain, from the University of Iowa in Iowa City, points out that statistical **wrangling** has been known to create mass extinctions where none occurred. It is easy to misinterpret data. For example, changes in sea level or inconsistent sampling methods can **mimic** major changes in biodiversity. Indeed, a recent and thorough examination of the literature on **marine bivalve** fossils has convinced David Jablonsky from the University of Chicago and his **colleagues** that their diversity has increased steadily over the past five million years.

H Adrain believes that fancy **analytical** techniques are no **substitute** for hard evidence, but he has also seen how **inadequate** historical collections can be. When he started his ongoing study of North American fossils from the Early Ordovician, around 500 million years ago, the literature described one genus and four species of **trilobites**, **lust** by going back to the fossil beds and sampling more thoroughly, Adrain found 11 genera and 39 species. "Looking inward has maybe taken us as far as it's going to take us," he says. "There's an **awful** lot more out there than is in the historical record." The only way to really get at the history of biodiversity, say Adrain and an increasingly vocal group of scientists, is to get back out in the field and collect new data.

I With an inventory of all living species, ecologists could start to put the current biodiversity crisis in historical perspective. Although creating such a list would be a task to **rival** even the Palaeobiology Database, it is exactly what the San Francisco-based ALL Species Foundation hopes to achieve in the next 25 years. The effort is essential, says Harvard biologist Edward O. Wilson, who is **alarmed** by current rates of extinction. "There is a crisis. We've begun to measure it, and it's very high," Wilson says, "We need this kind of information in much more detail to protect all of biodiversity, not just the ones we know well." Let the counting continue.

fungi　n.（fungus 的复数）真菌；霉，霉菌
impress　v. 印；给……以深刻印象
n. 印象；印记
mathematical　adj. 数学的；精确的；绝对的；可能性极小的
wizardry　n. 魔法；魔术；杰出能力
wrangle　v. 争吵；争论
mimic　v. 摹拟；模仿
n. 巧于模仿的人；复写品
adj. 模仿的，摹拟的
marine　adj. 海的；海产的；海军的
n. 水兵；海军士兵；海事，海运业
bivalve　n. 双壳类；双阀
colleague　n. 同事；同行
analytical　adj. 分析的，分析法的
substitute　v. 代替，替换
n. 替代物；代替者；替补
inadequate　adj. 不充足的；不适当的；不足胜任的
trilobite　n. 三叶虫
lust　n. 欲望；渴求
v. 贪求；渴望
awful　adj. 可怕的；糟糕的；非常的
rival　n. 对手；竞争者
v. 竞争；比得上
alarm　n. 警报；惊恐
v. 警告；使惊恐

Questions 1—6

Passage B has nine paragraphs, Paragraphs A—I.

Choose the correct heading for Paragraphs A—F from the list of headings below.

Write the correct number i—vii in the box 1—6.

| 1 | | 2 | | 3 | | 4 | | 5 | | 6 | |

List of Headings

ⅰ. Potential error exists in the database
ⅱ. Supporter of database declared its value
ⅲ. The purpose of building paleobiology data
ⅳ. Reason why some species were not included in it
ⅴ. Inaccuracy of breed due to different names
ⅵ. Achievement of the Paleobiology Database
ⅶ. Criticism of waste of fund on the project

1	Paragraph **A**	Answer	
2	Paragraph **B**	Answer	
3	Paragraph **C**	Answer	
4	Paragraph **D**	Answer	
5	Paragraph **E**	Answer	
6	Paragraph **F**	Answer	

Questions 7－9

Use the information in the passage to match the people (listed A－D) with opinions or deeds below. Write the appropriate letters, A－D, in boxes 7－9.

| 7 | | 8 | | 9 | |

A. Jonathan Adrain　　B. John Alroy
C. David Jablonsky　　D. Edward O. Wilson

7. _____ reckoned utilization of the database would help scientists learn patterns of diversity of extinction.

8. _____ believed contribution of detailed statics should cover the known species.

9. _____ reached a contradictory finding to the tremendous species die-out.

Questions 10 and 11

Choose the TWO correct letters from the following.

	10		11	

A. Almost all the experts welcome the project.

B. Intrigues both positive and negative opinions from various experts.

C. All different creatures in the database have unique names.

D. Aims to embrace all fossil information globally.

E. Gets more information from the record rather than the field.

Questions 12 and 13

Choose the correct letters A, B, C, or D.

	12		13	

12. According to the passage, jellyfish belongs to which category of the Palaeobiology Database?

   A. Repetition breed

   B. Untraceable species

   C. Specifically detailed species

   D. Currently living creature

13. What is the main idea of the last paragraph?

   A. Continue to complete missing species in the Palaeobiology Database.

   B. Stop to contributing to the Palaeobiology Database.

   C. Importance of create a database of living creature.

   D. Study more in the field rather than in the book.

Unit 3
Plate Tectonics and Earth Structure

 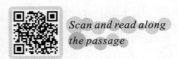

Scan and read along the passage

Plate Tectonics is the first theory to provide a comprehensive view of the processes that produced Earth's major surface features, including the continents and ocean basins. Prior to the 1960s, most geologists held the view that the ocean basins and continents had fixed geographic positions and were of great antiquity. Less than a decade later researchers came to realise that Earth's continents are not static, instead, they gradually migrate across the globe. Because of these movements, blocks of continental material collide, deforming the intervening crust, thereby creating Earth's great mountain chains. Furthermore, landmasses occasionally split apart. As the continental blocks separate, a new ocean basin emerges between them. Meanwhile, other portions of the seafloor plunge into the mantle. In short, a dramatically different model of Earth's tectonic processes emerged. In this unit, we will examine the events that led to this dramatic reversal of scientific opinion in an attempt to provide insight into how science works. We will briefly trace the development of the Continental Drift Hypothesis, examine why it was first rejected, and consider the evidence that finally led to the acceptance of its direct descendant—The Theory of Plate Tectonics.

Passage A

Continental Drift: an Idea Before Its Time[①]

Scan and read along the vocabulary

A The idea that continents, particularly South America and Africa, fit together like pieces of a jigsaw puzzle came about during the 1600s as better world maps became available. However, little significance was given to this notion until 1915, when Alfred

---

[①] This Passage is adapted from: Lutgens F K, Tarbuck E J. Plate tectonics: a scientific revolution unfolds[M]//Essentials of geology. 11 th ed. Upper Saddle River, NJ, USA: Pearson, 2012: 361-391.

Wegener[1] (1880—1930), a German **meteorologist** and **geophysicist**, wrote *The Origin of Continents and Oceans*. This book, published in several editions, set forth the basic outline of Wegener's **hypothesis** called Continental Drift—which dared to challenge the long-held assumption that the continents and ocean basins had fixed geographic positions.

B Wegener suggested that a single **supercontinent** consisting of all Earth's **landmasses** once existed. He named this giant landmass **Pangaea**. Wegener further hypothesised that about 200 million years ago, during the early part of the Mesozoic Era[2], this supercontinent began to fragment into smaller landmasses. These continental blocks then "drifted" to their present positions over a span of millions of years. Wegener and others who advocated the Continental Drift Hypothesis collected substantial evidence to support their point of view.

Evidence: fossils match across the seas

C Besides the puzzlelike fit of the continents, fossils provided support for continental drift. Fossils of the **reptile** Mesosaurus[3] have been found in South America and Africa. This swimming reptile lived in freshwater and on land. How could fossils of Mesosaurus be found on land areas separated by a large ocean of salt water? It probably could not swim between the continents. Wegener hypothesised that this reptile lived on both continents when they were joined. Another fossil that supports the hypothesis of continental drift is **Glossopteris**, the fossil of which has been found in Africa, Australia, India, South America, and Antarctica. The presence of Glossopteris in so many areas also supported Wegener's idea that all of these regions once were connected and had similar climates.

Evidence: rock types and geologic features

D Anyone who has worked a jigsaw puzzle knows that its successful completion requires that you fit the pieces together while maintaining the **continuity** of the picture. The "picture" that must match in the "continental drift puzzle" is one of rock types and geologic features such as mountain belts. If the continents were once together, the rocks found in a particular region on one continent should

---

meteorologist　*n.* 气象学者
geophysicist　*n.* 地球物理学者
hypothesis　*n.* 假设；
v. hypothesise〈英〉/hypothesize
〈美〉假定；假设；猜测
supercontinent　*n.* 超大陆
landmass　*n.* 大陆
Pangaea　*n.* all lands 泛大陆；盘古大陆
reptile　*adj.* 爬虫类的
n. 爬行动物
Glossopteris　*n.* 舌羊齿；舌蕨属
continuity　*n.* 连续性；一连串

closely match in age and type of those found in adjacent positions on the once **adjoining** continent. Wegener found evidence of 2.2-billion-year-old igneous rocks in Brazil that closely resembled similarly aged rocks in Africa. Similar evidence can be found in mountain belts that terminate at one coastline, only to reappear on landmasses across the ocean. For instance, the mountain belt that includes **the Appalachians** trends northeastward through the eastern United States and disappears off the coast of **Newfoundland**. Mountains of comparable age and structure are found in the British Isles[4], and Scandinavia[5]. When these landmasses are reassembled, the mountain chains form a nearly continuous belt.

Evidence: ancient climates

E Because Wegener was a student of world climates, he suspected that **paleoclimatic** data might also support the idea of mobile continents. His assertion was bolstered when he learned that evidence for a glacial period that dated to the late **Paleozoic** had been discovered in Southern Africa, South America, Australia, and India. This meant that about 300 million years ago, vast ice sheets covered extensive portions of the Southern **Hemisphere** as well as India. Much of the land area that contains evidence of this period of **Paleozoic glaciation** presently lies within 30 degrees of the equator in subtropical or tropical climates. Wegener suggested that these continents were connected and partly covered with ice near Earth's South Pole long ago. This would account for the conditions necessary to generate extensive expanses of glacial ice over much of these landmasses. At the same time, this geography would place today's northern continents nearer the equator and account for the tropical **swamps** that generated the vast **coal deposits**. Wegener was so convinced that his explanation was correct and wrote, "This evidence is so **compelling** that by comparison all other criteria must take a back seat."

F Although Wegener provided evidence to support his Hypothesis of Continental Drift, he could not explain how, when, or why these changes took place. Given that his idea was so radically different at that time, most people closed their minds to it. The respected American geologist Rollin Thomas Chamberlaineven commented that, "Wegener's hypothesis in general is of the foot-

---

adjoining *adj.* 邻接的；毗连的
v. adjoin 邻接
the Appalachians *n.* 美洲东北部阿巴拉契亚山脉
Newfoundland *n.* 纽芬兰（加拿大东部岛亦为加拿大省）
paleoclimatic *adj.* 古气候的（paleo＝acient, climatic＝climate）
Paleozoic *adj.* 古生代的
hemisphere *n.* 半球
Paleozoic glaciation *n.* 古生代冰期
swamp *n.* 湿地；沼泽
coal deposits *n.* 煤储量
compelling *adj.* 引人入胜的，扣人心弦的，非常强烈的，不可抗拒的
v. compel 强迫；使不得不

loose type, in that it takes considerable liberty with our globe, and is less bound by restrictions or tied down by awkward, ugly facts than most of its rival theories. Its appeal seems to lie in the fact that it plays a game in which there are few restrictive rules and no sharply drawn code of conduct." One main objection to Wegener's hypothesis stemmed from his inability to identify a credible mechanism for continental drift. Wegener proposed that gravitational forces of the Moon and Sun that produce Earth's tides were also capable of gradually moving the continents across the globe. However, the prominent physicist Harold Jeffreys correctly countered that tidal forces of the magnitude needed to displace the continents would bring Earth's rotation to a halt in a matter of a few years. Wegener also incorrectly suggested that the larger and sturdier continents broke through thinner oceanic crust, much like ice breakers cut through ice. However, no evidence existed to suggest that the ocean floor was weak enough to permit passage of the continents with out the continents being appreciably **deformed** in the process. In 1930, Wegener made his fourth and final trip to the Greenland ice sheet. Although the primary focus of this **expedition** was to study the harsh winter polar climate on the ice-covered island, Wegener continued to test his Continental Drift Hypothesis. As in earlier expeditions, he used **astronomical** methods in an attempt to verify that Greenland had drifted westward with respect to Europe. While returning from Eismitte (an experimental station located in the centre of Greenland), Wegener perished along with a companion. His intriguing idea, however, did not die. After Wegener's death, more clues were found, largely because of advances in technology and the development of new ideas related to continental drift, which ultimately contributed to the formation of The Theory of Plate Tectonics[6].

deform v. 变形；使成畸形
expedition n. 远征；探险队；迅速
astronomical adj. 天文的，天文学的；极大的

Notes

(1) Alfred Wegener (1880—1930)：德国气象学家和地球物理学家，最初于 1912 年提出大陆漂移观点，1915 年出版的 *The Origin of Continents and Oceans* 系统地论述了大陆漂移学说。

(2) the Mesozoic Era：中生代，是地质学专业术语。整个地壳历史划分为隐生宙和显生宙两大阶段。宙之下分代，隐生宙分为太古代、元古代，显生宙又划分为古生代、中生代、新生代。代之下又可划分若干纪，如中生代是显生宙(the Phanerozoic Eon)的 3 个地质时代之一，其下再分

为三叠纪、侏罗纪和白垩纪3个纪。中生代最早是由意大利地质学家Giovanni Arduino所建立,当时名为第二纪(Secondary),以相对于现今的第三纪。中生代介于古生代与新生代之间。由于这段时期的优势动物是爬行动物,尤其是恐龙,因此又称为爬行动物时代(Age of the Reptiles)。

(3) Mesosaurus:中龙。中龙是最早下水的爬行动物。主要生活在溪流和水潭中,很少上岸,特别爱吃水里的鱼。它身体细长,肩部和腰部的骨骼都比较小,身后有一条长而灵活的尾巴,脚较大,称为桡足,主要用尾巴游泳。它的上下颌特别长,嘴里长满锋利的牙齿,很适合捕鱼。中龙分别在非洲和南美洲发现,当时科学家认为这种小型的爬行动物是生活在淡水水域里的,不可能游过广阔的海洋而在南非和南美洲之间迁徙或散布,因此认定它是大陆漂移的有力证据。

(4) British Isles:不列颠群岛,包括大不列颠岛、爱尔兰岛、马恩岛及附近的5 500多个小岛,群岛上有英国和爱尔兰共和国。

(5) Scandinavia:斯堪的纳维亚(半岛),位于欧洲西北角,挪威海和波罗的海之间,是欧洲最大的半岛,也是世界第五大半岛,欧洲最大的半岛,面积约80万 km^2。半岛上有瑞典、挪威两国和芬兰一角。斯堪的纳维亚山脉(Scandinavian Mountains)横亘于两个国家之间,半岛西部属山地,沿岸陡峭,多岛屿和峡湾;东、南部地势较平整,半岛气候属温带气候,其北端严寒。

(6) The Theory of Plate Tectonics:板块构造论。

Online Resources

1. The Mesozoic [mesəuzəuɪk] **Era**(中生代) is an interval of geological time from about 252－66 million years ago. Mesozoic means "middle life", deriving from the Greek prefix "meso-/μεσο-" for "between" and "zōon/ζῶον" meaning "animal" or "living being". It is one of three geologic eras of the Phanerozoic Eon (显生宙), preceded by the Paleozoic Era(古生代) and succeeded by the Cenozoic Era(新生代). The era is subdivided into three major periods: the Triassic (三叠纪), Jurassic (侏罗纪), and Cretaceous (白垩纪), which are further subdivided into a number of epochs and stages. For more information about geologic times, please visit either the Chinese website (http://baike.sogou.com/h43696.htm? sp=l67150470), or the English one (https://en.wikipedia.org/wiki/Geologic_time_scale).

2. Glossopteris (舌羊齿;舌蕨属)(meaning "tongue", because the leaves are tongue-shaped) is the largest and best-known genus of the extinct order of seed ferns known as Glossopteridales (also known as Arberiales or Ottokariales). The genus Glossopteris refers only to leaves, within a framework of form genera used in paleobotany. For likely reproductive organs see Glossopteridaceae, and these are important because they indicate biological identity of these plants that were critical for recognising former connections between the varied fragments of Gondwana: South America, Africa, India, Australia, New Zealand, and Antarctica. (Retrieved from Wikipedia, https://en.wikipedia.org/wiki/Glossopteris.)

Unit 3　Plate Tectonics and Earth Structure

3. **The Appalachians**（阿巴拉契亚山脉）are a system of mountains in Eastern North America. The Appalachians first formed roughly 480 million years ago during the Ordovician Period. It once reached elevations similar to those of the Alps and the Rocky Mountains before naturally occurring erosion. The Appalachian chain is a barrier to east-west travel, as it forms a series of alternating ridgelines and valleys oriented in opposition to most roads running east or west. (Retrieved from Wikipedia, https://en.wikipedia.org/wiki/Appalachian_Mountains.)

Exercises

Detailed Understanding

Ⅰ. **Answer the following questions according to the passage you have read.**

1. What was the first line of evidence that led early investigators to suspect the continent were once connected?
2. What is Wegener's Continent Drift Hypothesis mainly about?
3. What evidence was used to support the Continental Drift Hypothesis?
4. Why does the discovery of the fossil remains of Mesosaurus and Glossopteris support the past existence of Pangaea?
5. How rock clues were used to support the Continental Drift Hypothesis?
6. How did Wegener account for the existence of glaciers in the southern landmasses at a time when areas in North America, Europe, and Asia supported lush tropical swamps?
7. To which two aspects of Wegner's Continental Drift Hypothesis did most Earth scientists object?

Vocabulary

Ⅱ. **Fill in the blanks with the words given below. Change the form where necessary.（15 个单词，10 个空）**

oppose	adjacent	outline	fossil	evidence
advocate	deposit	landmass	fragment	deposit
convince	hypothesis	support	assert	considerable

1. Wegener suggested that all continents once were connected as one large _____, that broke apart about 200 million years ago.
2. Most of the scientific community, particularly in North America, either categorically rejected continental drift or at least treated it with _____ skepticism.
3. Although many of Wegener's contemporaries his views _____, even to the point of open ridicule, some considered his ideas plausible.
4. The evidence Wegener presented had not been enough to _____ many people during his lifetime. He was unable to explain exactly how the continents drifted apart.

5. A classic example is Mesosaurus, an aquatic fish-catching reptile whose _____ remains are limited to black shales of the Permian period in Eastern South America and Southwestern Africa.
6. Since the remains of Mesosaurus have been found in both Africa and South America, Wegener _____ that South America and Africa must have been joined during that period of Earth history.
7. Wegener also cited the distribution of the fossil of Glossopteris as _____ for the existence of Pangaea.
8. Continental drift was particularly distasteful to North American geologists, perhaps because much of the _____ evidence had been gathered from the continents of Africa, South America, and Australia, with which most North American geologists were unfamiliar.
9. The fossils of warm-weather plants were found on the island of Spitsbergen in the Arctic Ocean. To explain this, Wegener _____ that Spitsbergen drifted from tropical regions to the Arctic.
10. How could you explain why glacial _____ are found in areas where no glaciers exist today?

Sentence Structure

Ⅲ. Combine the following simple sentences into one with proper connectives to make them more concise and clear.

> **Model**
> During the Ice Age, the lowering of sea level allowed mammals to cross the narrow Bering Strait.
> The Ice Age ended about 8000 years ago.
> The narrow Bering Strait separates Russia and Alaska.
> ⟶During the Ice Age that ended about 8000 years ago, the lowering of sea level allowed mammals to cross the narrow Bering Strait which separates Russia and Alaska.

1. A supercontinent began breaking apart about 200 million years ago.
   The supercontinent was called Pangaea.
   This was one of the major tenets of the Continental Drift Hypothesis.

2. Wegener contended, if land bridges of this magnitude once existed, their remains would still lie below sea level.
   Modern maps of the seafloor substantiate Wegener's contention.

3. Ol Doinyo Lengai is an active volcano in the East African Rift Valley.
   The East African Rift Valley is a place.
   In the place, Earth's crust is being pulled apart.

4. How did the continents drift apart?
   Wegener was unable to explain exactly.

5. Wegener intended to add credibility to his argument.
   Wegener documented cases of several fossil organisms.
   The fossil organisms were found on different landmasses.
   But the landmasses are presently separated by the vast ocean.
   It was unlikely possible that the living forms of the fossil organisms could have crossed the vast ocean.

Translation
IV. Translate the following sentences into Chinese.
1. The idea that continents, particularly South America and Africa, fit together like pieces of a jigsaw puzzle came about during the 1600s as better world maps became available.

2. If the continents were once together, the rocks found in a particular region on one continent should closely match in age and type of those found in adjacent positions on the once adjoining continent.

3. For instance, the mountain belt that includes the Appalachians trends northeastward through the eastern United States and disappears off the coast of Newfoundland.

4. The respected American geologist Rollin Thomas Chamberlain even commented that, "Wegener's hypothesis in general is of the foot-loose type, in that it takes considerable liberty with our globe, and is less bound by restrictions or tied down by awkward, ugly facts than most of its rival theories. Its appeal seems to lie in the fact that it plays a game in which there are few restrictive rules and no sharply drawn code of conduct."

5. After Wegener's death, more clues were found, largely because of advances in technology and the development of new ideas related to continental drift, which ultimately contributed to the formation of the Theory of Plate Tectonics.

Ⅴ. **Translate the following paragraph into English.**

大陆漂移学说于1915年由德国地质学家兼气象学家阿尔弗雷德·魏格纳首次提出。魏格纳认为两亿年前存在一个巨大的超大陆,他把它称之为"泛大陆"或"盘古大陆",寓以"所有的土地"之意。该泛大陆从侏罗纪时期开始,先分化成劳亚古大陆和冈瓦纳大陆两个稍小的超大陆。直到白垩纪末期,地球陆地地块才分化成现今的模样。魏格纳将该学说于1915年发表于专著《海陆的起源》一书中。在书中,魏格纳提出,部分地壳缓慢漂移在液芯上,并援引化石记录、岩石形态和冰川沉积作为支撑证据。

Reading Skills

Signpost Language[①]

In academic writing, the author is responsible for making the text as clear as possible for the reader. To achieve clarity, the author has to ensure that the writing is explicit. Signpost language, words or phrases that express a connection between two ideas or make the transition from one point to the next, can enhance the clarity of your argument by pointing out the relationship between your ideas, and where your line of thought is going. The simple adding of a few signposts can make the writing much more readable.

> Incorporation offers several advantages to businesses and their owners. **For one thing**, ownership is easy to transfer. The business is able to maintain a continuous existence even when the original owners are no longer involved (Brown, 1999). **In addition**, the stock holders of a corporation are not held responsible for the business's debts (Henry, 2009). If the XYZ Corporation defaults on a $1 million loan, **for instance**, its investors will not be held responsible for paying that liability. Incorporation **also** enables a business to obtain professional managers with centralised authority and responsibility; **therefore**, the business can be run more efficiently (Schwartz, 2010). **Finally**, incorporation gives a business certain legal rights. **For example**, it can enter into contracts, own property, and borrow money (Brown, 1999).

Transitions show the reader the "movement" between ideas/points. They show that the ideas follow a logical order and build on each other, creating "flow". If a paragraph flows well from point to point, it should be obvious to the reader when you move from one point to another.

They tell the reader:
(1) how the main ideas support the thesis statement.
(2) how each group of ideas follow from the ones before, and whether information is
- an additional point.
- in contrast to what has been said.
- an example.

For example, if you are analysing one study and then comparing it to another in a later paragraph, a transition word or phrase could highlight the change in direction or the creation of a comparison:

> **In contrast to** the conclusion drawn by Smith (2004), Nguyen (2006) showed that the connection between the factors was not causal in most circumstances.

[①] This section is adapted from: ⓐ Newcastle University. Signposting [EB/OL]. [2017-08-15]. http://www.ncl.ac.uk/students/wdc/learning/language/signposting.htm. ⓑ Massey University. Signpost words and phrases [EB/OL]. [2017-05-20]. http://owll.massey.ac.nz/pdf/studyup-essays-2-handout.pdf.

A paragraph that provides a similar point to a previous one could start as follows:

> **Similarly,** Bell (2006) highlights that …

Signpost language can be briefly grouped into two categories:

- **major signposts** that signal key aspects of the work, such as purpose, structure, author's stance, main points, directions of the argument, conclusions.
- **linking words and phrases** that show connections between sentences and paragraphs.

Some of the most common signposts are listed hereinafter:

Examples of major signposts

The aim of the study is to …

The purpose of the thesis is to …

This essay argues that …

The main questions addressed in this paper are …

This essay critically examines …

The above discussion raises some interesting questions.

The paper begins by… It will then go on to … Finally …

This chapter reviews the literature …

In conclusion, …

Examples of linking words and phrases

(1) *Highlighting or emphasising a point*:

Importantly, …

Indeed, …

In fact, …

More importantly, …

Furthermore, …

Moreover, …

It is also important to highlight …

Adding a similar point, …

Similarly, …

Likewise, …

Again, …

Also, …

(2) *Changing direction or creating a comparison*:

However, …

Rather, …

In contrast, …

Conversely, …

On one hand, …

On the other hand, ...
   In comparison, ...
   Compared to ...
   Another point to consider is ...
(3) *Summarising:*
   Finally, ...
   Lastly, ...
   In conclusion, ...
   To summarise, ...
   In summary, ...
   Overall, ...
   The three main points are ...
(4) *Being more specific:*
   In particular, ...
   In relation to ...
   More specifically, ...
   With respect to ...
   In terms of ...
(5) *Acknowledging something and moving to a different point:*
   Although ...
   Even though ...
   Despite ...
   Notwithstanding ...
(6) *Giving an example:*
   For instance, ... / For example, ...
   This can be illustrated by ...
   ... , namely, ...
   ... , such as ...
(7) *Following a line of reasoning:*
   Therefore, ...
   Subsequently, ...
   Hence, ...
   Consequently, ...
   Accordingly, ...
   As a result, ...
   As a consequence, ...
   To this end, ...

Passage B

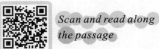

Plate Tectonics[①]

A   The new development of oceanography in the 1960s advanced Wegener's Continental Drift Hypothesis. Following World War Ⅱ, oceanographers equipped with new marine tools and ample funding from the U. S. Office of Naval Research **embarked on** an **unprecedented** period of oceanographic exploration. Over the next two decades a much better picture of large expanses of the seafloor slowly emerged. From this work came the discovery of a global **oceanic ridge** system that winds through all of the major oceans in a manner similar to the seams on a baseball. In other parts of the ocean, new discoveries were also made. Earthquake studies conducted in the western Pacific demonstrated that **tectonic** activity was occurring at great depths beneath deep-ocean **trenches**. Of equal importance was the fact that dredging of the seafloor did not bring up any oceanic crust that was older than 180 million years. Further, **sediment** accumulations in the deep-ocean basins were found to be thin, not the thousands of metres that were predicted.

B   By 1968, these developments, among others, led to the unfolding of a far more encompassing theory than continental drift, known as **plate tectonics**. According to the theory of plate tectonics, Earth's crust and part of the upper mantle are broken into sections. These sections, called plates, move on a plasticlike layer of the mantle. The plates can be thought of as rafts that float and move on this layer. They are made of the crust and a part of the upper mantle. These two parts combined are the **lithosphere**. This rigid layer is about 100km thick and generally is less dense than material underneath. The plasticlike layer below the lithosphere is called the **asthenosphere**. The rigid plates of the lithosphere float and move around on the asthenosphere.

C   Because plates are in constant motion, they unavoidably interact with one another. Most major interactions among them

---

[①] This passage is adapted from: Feather R M, Jr. , Snyder S L, Zike D. Plate tectonics[M]//Earth Science. New York, NY, USA: McGraw-Hill, 2005: 270 – 297.

(and, therefore, most deformation) occur along their **boundaries**. In fact, plate boundaries were first established by plotting the locations of earthquakes and volcanoes. Plates are bounded by three distinct types of boundaries, which are **differentiated** by the type of movement they exhibit. Divergent **plate boundaries** occur where plates move apart, resulting in **upwelling** of material from the mantle to create new seafloor. Most divergent boundaries occur along the axis of the oceanic ridge system and are **associated with seafloor spreading**. New divergent boundaries may form within a continental (for example, the East African Rift Valleys), where they may fragment a landmass and develop a new ocean basin. Convergent plate boundaries occur where plates move together, resulting in the **subduction** of oceanic lithosphere into the mantle along a deep-ocean trench. Convergence of an oceanic and continental block results in subduction of the oceanic slab and the formation of a **continental volcanic arc**, such as **the Andes of South America**. Oceanic-oceanic convergence results in an arc-shaped chain of volcanic islands called a **volcanic island arc**. When two plates carrying continental crust converge, the **buoyant** continental blocks collide, resulting in the formation of a mountain belt as exemplified by the Himalayas. Transform fault boundaries occur where plates **grind** past each other without the production or destruction of lithosphere. Most transform faults join two segments of a mid-ocean ridge where they provide the means by which oceanic crust created at a **ridge crest** can be transported to its site of destruction—a deep-ocean trench. Still others, like **the San Andreas Fault**, cut through continental crust.

D Wegener was unable to come up with an explanation for why plates move. Today, researchers tend to adopt convection as possible explanations. They suggest that it is the transfer of heat inside Earth that provides the energy to move plates and causes many of Earth's surface features. They argue that Earth is a dynamic planet with a hot interior. This heat leads to convection, which powers the movement of plates. As the plates move, they interact. The interaction of plates produces forces that build mountains, create ocean basins, and cause volcanoes. When rocks in Earth's crust break and move, energy is **released** in the form of seismic waves. Humans feel this release as earthquakes. They can

boundary *n.* 边界；范围；分界线
differentiate *vi.* 区分，区别
vt. 区分，区别
plate boundary *n.* 板块边界
upwell *vi.* 上涌；往上涌出
associate with *v.* 联合；与……联系在一起；和……来往
seafloor spreading *n.* 海底扩张
subduction *n.* 俯冲；除去；减法
continental volcanic arc *n.* 火山弧
the Andes of South America *n.* 南美安第斯山脉
volcanic island arc *n.* 火山岛弧，大洋中呈弧形分布且有火山活动的群岛
buoyant *adj.* 轻快的；有浮力的；上涨的
grind *vt.* 磨碎；磨快
vi. 磨碎；折磨
ridge crest *n.* 脊顶
the San Andreas Fault *n.*（美国）圣安德列斯断层
release *vt.* 释放；发射；让与；允许发表
n. 释放；发布；让与

also see some effects of plate tectonics in mountain regions, where volcanoes erupt or where landscapes have been changed from past earthquakes or volcanic activities.

E Tension forces, or forces that pull apart, can stretch Earth's crust, causing large blocks of crust to break and **tilt** or slide down the broken surfaces of crust. When rocks break and move along surfaces, a **fault** forms. Faults interrupt rock layers by moving them out of place. Entire mountain ranges can form in the process, which are often called **fault-block mountains**. Generally, the faults that form from pull-apart forces are normal faults—faults in which the rock layers above the fault move down when compared with rock layers below the fault. Rift valleys and mid-ocean ridges can form where Earth's crust separates. Examples of rift valleys are **the Great Rift Valley** in Africa, and the valleys that occur in the middle of mid-ocean ridges. Examples of mid-ocean ridges include **the Mid-Atlantic Ridge** and **the East Pacific Rise.**

F Compression forces, by contrast, can squeeze objects together. Where plates come together, compression forces produce several effects. As continental plates collide, the forces that are generated cause massive **folding** and faulting of rock layers into mountain ranges such as the Himalaya or the Appalachian Mountains. The type of faulting produced is generally **reverse faulting**. Along a reverse fault, the rock layers above the fault surface move up relative to the rock layers below the fault.

G At transform boundaries, two plates slide past one another without converging or diverging. The plates stick and then slide, mostly in a horizontal direction, along large **strike-slip faults**. In a strike-slip fault, rocks on opposite sides of the fault move in opposite directions, or in the same direction at different rates. One such example is the San Andreas Fault. When plates move suddenly, **vibrations** are generated inside Earth that are felt as an earthquake.

H With the development of the theory of plate tectonics, researchers began testing this new model of how Earth works. But most tests scientists could use to check for plate movement were indirect. They could study the magnetic characteristics of rocks on the seafloor. They could study volcanoes and earthquakes. These methods supported the theory that the plates have moved and are

tilt *vi.* 倾斜；翘起；以言词或文字抨击
vt. 使倾斜；使翘起
n. 倾斜
fault *n.* 故障；错误；缺点；毛病；〈地质〉断层（normal fault *n.* 正断层）
vi. 弄错；产生断层
fault-block mountains
n. 断块山地
fault-fold mountains
n. 断褶山地
the Great Rift Valley *n.* 东非大裂谷
the Mid-Atlantic Ridge *n.* 大西洋中脊
the East Pacific Rise *n.* 东太平洋海隆（东太平洋海岭）
fold *n.* 褶皱
reverse fault *n.* 逆断层；逆向断层
strike-slip fault *n.* 平移断层；走滑断层
vibration *n.* 振动

still moving. However, they did not provide proof—only support—of the idea. New methods have to be discovered to be able to measure the small amounts of movement of Earth's plates. Although there is still much to be learned about the **mechanisms** that cause Earth's tectonic plates to migrate across the globe, one thing is clear. The unequal distribution of heat in Earth's interior gene-rates some type of **thermal** convection that ultimately drives plate-mantle motion.

mechanism　*n.* 机制；原理，途径；机械装置；技巧
thermal　*adj.* 热的；热量的；保热的

Questions 1—4

Passage B has eight paragraphs, A—H.

Choose the correct heading for paragraphs B—D and H from the list of headings below.

Write the correct number i—xi in boxes 1—4.

| 1 | | 2 | | 3 | | 4 | |

List of Headings

i. Plate tectonics in the future
ii. Convergent boundaries
iii. Convection inside Earth
iv. Causes of plate tectonics
v. Measuring plate motion
vi. The development of the plate tectonics theory
vii. Features caused by compression forces of plates interaction
viii. The main tenets of the plate tectonics theory
ix. Features caused at transform boundaries
x. Features caused by tension forces of plates interaction
xi. Three types of plate boundaries

Example	Paragraph **A**	Answer	vi
1	Paragraph **B**	Answer	
2	Paragraph **C**	Answer	
3	Paragraph **D**	Answer	
Example	Paragraph **E**	Answer	x
Example	Paragraph **F**	Answer	vii
Example	Paragraph **G**	Answer	ix
4	Paragraph **H**	Answer	

Questions 5—10

Do the following statements reflect the claims of the writer in Passage B?

In boxes 5—10, write

    Y(YES)   if the statement reflects the claims of the writer
    N(NO)   if the statement contradicts the claims of the writer
    NG(NOT GIVEN)   if it is impossible to say what the writer thinks about this

5	6	7	8	9	10

5. According to plate tectonics, Earth's rigid outer layer (lithosphere) overlies a weaker region called the asthenosphere.
6. Many of the topical islands in the Caribbean, where Americans dream of taking winter holidays, are of volcanic origin.
7. Tension forces cause normal faults, rift valleys, and mid-ocean ridges at divergent boundaries.
8. Divergent boundaries occur where two plates move apart, resulting in oceanic lithosphere descending beneath an overriding plate, eventually to be reabsorbed into the mantle.
9. A few geologists speculate that the next supercontinent may form as a result of subduction of the floor of the Atlantic Ocean, resulting in the collision of the Americas with the Eurasian-African landmass.
10. At convergent boundaries, compression forces cause folding, strike-slip faults, and mountains.

Questions 11—13

Complete the summary of Paragraphs B—G with the list of words A—H below.

Write the correct letter A—H in boxes 11—13.

11	12	13

    The theory of plate tectonics states that sections of the seafloor and continents move as plates on a plastic-like layer of the mantle. The boundary between two plates moving apart is called a divergent boundary. Plates move together at a convergent boundary. Transform boundaries occur where two plates __11__ past one another. __12__ is/are thought to cause the movement of Earth's plates. Tension forces cause normal faults, rift valleys, and mid-ocean ridges at __13__ boundaries. At convergent boundaries, compression forces cause folding, reverse faults, and mountains. At transform boundaries, two plates slide past one another along strike-slip faults.

| A. convergent | B. divergent | C. trench | D. plate subduction |
| E. slide | F. move | G. ocean basins | H. convection currents |

Unit 4
Mineral and Rock

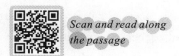

A mineral has one specific chemical composition, whereas a rock can be an aggregate of different minerals or mineraloids. The silicate minerals compose over 90% of Earth's crust. The study of minerals is called mineralogy. Earth's outer solid layer, the lithosphere, is made of rock. The scientific study of rocks is called petrology, which is an essential component of geology.

Passage A

Mineral[①]

A A mineral is a naturally occurring chemical compound, usually of **crystalline** form and **abiogenic** in origin. A mineral has one specific chemical composition, whereas a rock can be an **aggregate** of different minerals or **mineraloids**.

B One definition of a mineral **encompasses** the following criteria:
- naturally occurring;
- stable at room temperature;
- represented by a chemical **formula**;
- usually abiogenic (not resulting from the activity of living organisms);
- ordered atomic arrangement.

C The first three general characteristics are less debated than the last two. The first criterion means that a mineral has to form

crystalline *adj.* 水晶的;结晶质的
n. 结晶性,结晶度
abiogenic *adj.* 自然发生的
aggregate *n.* 合计;聚集体
mineraloids *n.* 准矿物;似矿物
encompass *vt.* 包含;包括
formula *n.* 公式;方案

---

① This passage is adapted from: Wikipedia. Mineral[EB/OL]. (2017-11-08)[2017-11-11]. https://en.wikipedia.org/wiki/Mineral.

by a natural process, which **excludes anthropogenic** compounds. Stability at room temperature[1], in the simplest sense, is **synonymous** to the mineral being solid. A mineral is an element or chemical compound that is normally crystalline and that has been formed as a result of geological processes.

D In addition, **biogenic** substances were explicitly excluded. Biogenic substances are chemical compounds produced entirely by biological processes without a geological component (e.g., **urinary calculi**, **oxalate** crystals in plant tissues, etc.) and are not regarded as minerals. However, if geological processes were involved in the **genesis** of the compound, then the product can be accepted as a mineral.

E Finally, the requirement of an ordered atomic arrangement is usually synonymous with crystallinity. However, crystals are also **periodic**, so the broader criterion is used instead. An ordered atomic arrangement gives rise to a variety of **macroscopic** physical properties, such as crystal structure and habit, hardness, **lustre**, color and streak.

F Classifying minerals ranges from simple to difficult. A mineral can be identified by several physical properties, some of them being sufficient for full identification without **equivocation**. In other cases, minerals can only be classified by more complex optical, chemical or X-ray **diffraction** analysis; these methods, however, can be costly and time-consuming. Physical properties applied for classification include crystal structure and habit, hardness, lustre and **diaphaneity**, colour and streak.

Crystal structure and habit

G Crystal structure results from the orderly **geometric** spatial arrangement of atoms in the internal structure of a mineral. This crystal structure is based on regular internal atomic or **ionic** arrangement that is often expressed in the geometric form that the crystal takes. Even when the mineral grains are too small to see or are irregularly shaped, the **underlying** crystal structure is always periodic and can be determined by X-ray diffraction. Minerals are typically described by their **symmetry** content. Crystals are restricted to 32 point groups[2], which differ by their symmetry. These groups are classified in turn into more broad categories, the

exclude *vt.* 排斥；排除，不包括
anthropogenic *adj.* 人为的，人类活动产生的
synonymous *adj.* 同义词的；同义的
biogenic *adj.* 源于生物的，生物所造成的
urinary *adj.* 尿的；泌尿的
calculi *n.* 结石；积石
oxalate *n.* 草酸盐
genesis *n.* 创始，起源，发生
periodic *adj.* 周期的；定期的；间歇的
macroscopic *adj.* 宏观的；肉眼可见的
lustre *n.* 光泽；光彩
equivocation *n.* 模棱两可的话，含糊话
diffraction *n.* 衍射
diaphaneity *n.* 透明度；透明性
geometric *adj.* 几何学的
ionic *adj.* 离子的
underlying *adj.* 潜在的；基础的；表面下的
symmetry *n.* 对称；匀称

most encompassing of these being six crystal families.

H Crystal habit refers to the overall shape of crystal. Several terms are used to describe this property. Common habits include **acicular**, which described needlelike crystals like in **natrolite**, bladed, **dendritic** (tree-pattern, common in native copper), **equant**, which is typical of garnet, **prismatic** (**elongated** in one direction), and **tabular**, which differs from bladed habit in that the former is **platy** whereas the latter has a defined elongation. Related to crystal form, the quality of crystal faces is **diagnostic** of some minerals, especially with a **petrographic** microscope. **Euhedral** crystals have a defined external shape, while **anhedral** crystals do not; those intermediate forms are termed **subhedral**.

Hardness

I The hardness of a mineral[3] defines how much it can resist scratching. This physical property is controlled by the chemical composition and crystalline structure of a mineral. A mineral's hardness is not necessarily constant for all sides, which is a function of its structure; **crystallographic** weakness renders some directions softer than others. An example of this property exists in **kyanite**, which has a Mohs hardness of 5½ parallel to [001] but 7 parallel to [100].

J The most common scale of measurement is the ordinal Mohs hardness scale. Defined by ten indicators, a mineral with a higher index scratches those below it. The scale ranges from **talc**, a **phyllosilicate**, to diamond, a carbon **polymorph** that is the hardest natural material. The scale is provided below:

Mohs Hardness	Mineral	Chemical Formula
1	Talc	$Mg_3Si_4O_{10}(OH)_2$
2	**Gypsum**	$CaSO_4 \cdot 2H_2O$
3	**Calcite**	$CaCO_3$
4	**Fluorite**	CaF_2
5	**Apatite**	$Ca_5(PO_4)_3(OH,Cl,F)$
6	**Orthoclase**	$KAlSi_3O_8$
7	**Quartz**	SiO_2
8	**Topaz**	$Al_2SiO_4(OH,F)_2$
9	**Corundum**	Al_2O_3
10	Diamond	C

acicular *adj.* 针状的,针尖状的
natrolite *n.* 钠沸石
dendritic *adj.* 树枝状的
equant *adj.* 等径的,等分的
prismatic *adj.* 棱形;棱镜的
elongate *vt.* 延长,加长
tabular *adj.* 表格的;平坦的;平板的
platy *adj.* 裂成平坦薄片的,板状的
diagnostic *adj.* 诊断的,判断的;特征的
petrographic *adj.* 岩相学的,岩类学的
euhedral *adj.* 自形的
anhedral *n.* 上反角
subhedral *adj.* 半形的,仅有部分晶面的
crystallographic *adj.* 晶体的;晶体学的
kyanite *n.* 蓝晶石
talc *n.* 滑石,云母
phyllosilicate *n.* 页硅酸盐
polymorph *n.* 多晶型物;多形态的动(植)物
gypsum *n.* 石膏
calcite *n.* 方解石
fluorite *n.* 萤石,氟石
apatite *n.* 磷灰石
orthoclase *n.* 正长石
quartz *n.* 石英
topaz *n.* 黄玉,托帕石
corundum *n.* 刚玉,金刚砂

Lustre and diaphaneity

K Lustre indicates how light reflects from the mineral's surface, with regards to its quality and intensity. There are numerous **qualitative** terms used to describe this property, which are split into metallic and non-metallic categories. Metallic and sub-metallic minerals have high **reflectivity** like metal; examples of minerals with this lustre are **galena** and **pyrite**. Non-metallic lustres include: **adamantine**, such as in diamond; **vitreous**, which is a glassy lustre very common in **silicate** minerals; pearly, such as in talc and **apophyllite**, **resinous**, such as members of the **garnet** group, silky which common in **fibrous** minerals such as **asbestiform chrysotile**.

L The diaphaneity of a mineral[4] describes the ability of light to pass through it. Transparent minerals do not diminish the intensity of light passing through it. An example of such a mineral is **muscovite** (potassium mica); some varieties are sufficiently clear to have been used for windows. **Translucent** minerals allow some light to pass, but less than those that are **transparent Jadeite** and **nephrite** (mineral forms of jade are examples of minerals with this property). Minerals that do not allow light to pass are called **opaque**. The diaphaneity of a mineral depends on thickness of the sample. When a mineral is sufficiently thin (e.g., in a thin section for **petrography**), it may become transparent even if that property is not seen in hand sample. In contrast, some minerals, such as **hematite** or pyrite are opaque even in thin-section.

Colour and Streak

M Colour is the most obvious property of a mineral, but it is often non-diagnostic. It is caused by electromagnetic radiation interacting with electrons (except in the case of **incandescence**, which does not apply to minerals). Two broad classes of elements (**idiochromatic** and **allochromatic**) are defined with regards to their contribution to a mineral's colour. Idiochromatic elements are essential to a mineral's composition and their contribution to a mineral's colour is diagnostic. Examples of such minerals are **malachite** (green) and **azurite** (blue). In contrast, allochromatic elements in minerals are present in trace amounts as impurities. An example of such a mineral would be the **ruby** and **sapphire**

---

qualitative   *adj.* 定性的,定质的;性质上的
reflectivity   *n.* 反射率
galena   *n.* 方铅矿
pyrite   *n.* 黄铁矿
adamantine   *adj.* 非常坚硬的 *n.* 金刚合金
vitreous   *adj.* 玻璃(似)的,玻璃质的 *n.* 玻璃状态,透明性
silicate   *n.* 硅酸盐
apophyllite   *n.* 鱼眼石
resinous   *adj.* 树脂质的,含树脂的,用树脂做的
garnet   *n.* 石榴石
fibrous   *adj.* 含纤维的,纤维性的
asbestiform   *adj.* 石棉状的
chrysotile   *n.* 温石绒,温石棉
muscovite   *n.* 白云母
translucent   *adj.* 半透明的;透亮的,有光泽的
transparent   *adj.* 透明的;清澈的
jadeite   *n.* 硬玉;翠
nephrite   *n.* 软玉
opaque   *adj.* 不透明的;无光泽的,晦暗的
petrography   *n.* 岩石记述学
hematite   *n.* 赤铁矿
incandescence   *n.* 白炽;白热
idiochromatic   *adj.* 自色的;本质色的
allochromatic   *adj.* 无色但含有色杂质的矿物的
malachite   *n.* 孔雀石
azurite   *n.* 石青;蓝铜矿
ruby   *n.* 红宝石
sapphire   *n.* 蓝宝石

varieties of the mineral corundum. The colours of pseudochromatic minerals are the result of interference of light waves. Examples include **labradorite** and **bornite**.

labradorite　*n.* 拉长岩
bornite　*n.* 斑铜矿

Ⓝ　The streak of a mineral refers to the colour of a mineral in powdered form, which may or may not be identical to its body colour. The most common way of testing this property is done with a streak plate, which is made out of porcelain and coloured either white or black. The streak of a mineral is independent of trace elements or any weathering surface. A common example of this pro- perty is illustrated with hematite, which is coloured black, silver, or red in hand sample, but has a cherry-red to reddish-brown streak. Streak is more often distinctive for metallic minerals, in contrast to non-metallic minerals whose body colour is created by allochromatic elements. Streak testing is constrained by the hardness of the mineral, as those harder than seven powder the streak plate instead.

Notes

(1) stability at room temperature:室温下的稳定性,指矿物具有固态特点。
(2) 32 point groups:32个质点群。晶体是内部质点(原子、离子)在三维空间周期性重复排列(即有序排列)的固体。由于质点呈有序排列,晶体内部就具有格子构造,称为晶体结构。不同晶体,其质点种类不同,质点的排列方向和间距不同,因而具有不同的晶体结构。
(3) hardness of a mineral 矿物硬度。指矿物抵抗外来机械作用力(如刻划、压入、研磨等)侵入的能力。
(4) diaphaneity of a mineral:矿物的透明度。矿物的光学性质是指自然光作用于矿物表面之后所发生折射和吸收等一系列光学效应所表现出来的各种性质,包括矿物的颜色、条痕、透明度及光泽等。

Online Resources

Mohs Hardness Scale（莫氏硬度表）

The Mohs scale of mineral hardness is a qualitative ordinal scale characterizing scratch resistance of various minerals through the ability of harder material to scratch softer material. Created in 1812 by German geologist and mineralogist Friedrich Mohs, it is one of several definitions of hardness in materials science, some of which are more quantitative. For more information, please visit https://en.wikipedia.org/wiki/Mohs_scale_of_mineral_hardness.

Exercises

Detailed Understanding

Ⅰ. Answer the following questions according to the passage you have read.

1. Tell the difference between a mineral and a rock.
2. Explain the fourth criterion defining a mineral.
3. What physical properties does ordered atomic arrangement refer to?
4. What methods of classifying minerals can be expensive and time-consuming?
5. A mineral's hardness is not necessarily constant for all sides, which is a function of its structure; crystallographic weakness renders some directions softer than others. Please take an example to illustrate it.
6. Which one is the hardest natural material?
7. Metallic and sub-metallic minerals have high reflectivity like metal. Please take one or two examples.
8. What does diaphaneity of a mineral depend on?

Vocabulary

Ⅱ. Fill in the blanks with the words given below. Change the form where necessary. (15个单词, 10个空)

composition	aggregate	encompass	criterion	formula
organism	exclude	stability	synonymous	explicitly
periodic	luster	optical	underlying	streak

1. The rate of growth of GNP will depend upon the rate of growth of _____ demand.
2. His repertoire in the past 2 years _____ everything from Bach to Schoenberg.
3. Television has transformed the size and social _____ of the audience at great sporting occasions.
4. Each hotel is inspected and, if it fulfils certain _____ is recommended.
5. The academy _____ women from its classes.
6. Going grey is not necessarily _____ with growing old.
7. Tell them _____ that you want to continue the friendship.
8. To solve a problem you have to understand its _____ causes.
9. Peace and _____ in the world need the active involvement of China.
10. _____ checks are taken to ensure that high standards are maintained.

Unit 4　Mineral and Rock

Sentence Structure

Ⅲ. Combine each pair of the following sentences, using "whereas" to compare or contrast two facts.

> **Model**
> A mineral has one specific chemical composition.
> A rock can be an aggregate of different minerals or mineraloids.
> ⟶A mineral has one specific chemical composition, whereas a rock can be an aggregate of different minerals or mineraloids.

1. Pensions are linked to inflation.
   Pensions should be linked to the cost of living.

2. You eat a massive plate of food for lunch.
   I have just a sandwich.

3. He had never done anything for them.
   They had done everything for him.

4. One's life is finite.
   One's passion for life is infinite.

5. These gases trap Sun's heat.
   Sulphur dioxide cools the atmosphere.

Translation

IV. Translate the following sentences into Chinese.

1. A mineral is a naturally occurring chemical compound, usually of crystalline form and abiogenic in origin.

2. The first criterion means that a mineral has to form by a natural process, which excludes anthropogenic compounds.

3. An ordered atomic arrangement gives rise to a variety of macroscopic physical properties, such as crystal structure and habit, hardness, luster, color and streak.

4. A mineral can be identified by several physical properties, some of them being sufficient for full identification without equivocation.

5. Lustre indicates how light reflects from the mineral's surface, with regards to its quality and intensity.

V. Translate the following passage into English.

中国现已发现171种矿产资源,查明资源储量的有158种,其中石油、天然气、煤、铀、地热等能源矿产10种,铁、锰、铜、铝、铅、锌等金属矿产54种,石墨、磷、硫、钾盐等非金属矿产91种,地下水、矿泉水等水气矿产3种。矿产地近18 000处,其中大中型矿产地7000余处。

Unit 4 Mineral and Rock

Reading Skills

Examples as Contextual Clues

Good writers often provide examples as clues or illustrations for readers in order to make meanings clear. Hence, it is always an effective way to learn to find out the examples in the reading material.

1. A mineral's hardness is not necessarily constant for all sides, which is a function of its structure; crystallographic weakness renders some directions softer than others. *An example of* this property exists in kyanite, which has a Mohs hardness of 5½ parallel to [001] but 7 parallel to [100].

2. Metallic and sub-metallic minerals have high reflectivity like metal; *examples of* minerals with this lustre are galena and pyrite.

3. Transparent minerals do not diminish the intensity of light passing through it. *An example of* such a mineral is muscovite (potassium mica); some varieties are sufficiently clear to have been used for windows.

4. Idiochromatic elements are essential to a mineral's composition and their contribution to a mineral's colour is diagnostic. *Examples of* such minerals are malachite (green) and azurite (blue).

5. In contrast, allochromatic elements in minerals are present in trace amounts as impurities. *An example of* such a mineral would be the ruby and sapphire varieties of the mineral corundum.

6. The streak of a mineral is independent of trace elements or any weatheringsurface. *A common example of* this property is illustrated with hematite, which is coloured black, silver, or red in hand sample, but has a cherry-red to reddish-brown streak.

7. Rock is a natural substance, a solid aggregate of one or more minerals or mineraloids. *For example*, granite, a common rock, is a combination of the minerals quartz, feldspar and biotite.

8. Plutonic or intrusive rocks result when magma cools and crystallises slowly within Earth's crust. *A common example of* this type is granite.

9. A gneiss has visible bands of differing lightness, *with a common example* being the granite gneiss.

10. Other varieties of foliated rock include slates, phyllites, and mylonite. *Familiar examples of* non-foliated metamorphic rocks include marble and soapstone.

Passage B

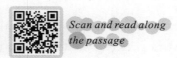
Scan and read along the passage

Classification of Rocks[①]

Scan and read along the vocabulary

granite　n. 花岗岩,花岗石
feldspar　n. 长石
biotite　n. 黑云母
granular　adj. 颗粒状的
permeability　n. 渗透性;磁导率

A　Rock is a natural substance, a solid aggregate of one or more minerals or mineraloids. For example, **granite**, a common rock, is a combination of the minerals quartz, **feldspar** and **biotite**. Earth's outer solid layer, the lithosphere, is made of rock. The scientific study of rocks is called petrology, which is an essential component of geology.

B　At a **granular** level, rocks are composed of grains of minerals, which, in turn, are homogeneous solids formed from a chemical compound that is arranged in an orderly manner. The aggregate minerals forming the rock are held together by chemical bonds. The types and abundance of minerals in a rock are determined by the manner in which the rock was formed. Many rocks contain silica (SiO_2); a compound of silicon and oxygen that forms 74.3% of Earth's crust. This material forms crystals with other compounds in the rock. The proportion of silica in rocks and minerals is a major factor in determining their name and properties. Rocks are geologically classified according to characteristics such as mineral and chemical composition, **permeability**, the texture of the constituent particles, and particle size. These physical properties are the end result of the processes that formed the rocks. Over the

---

[①]This passage is adapted from: Wikipedia. Rocks[EB/OL]. (2017-12-15)[2017-12-25]. https://en.wikipedia.org/wiki/Rock_(geology).

course of time, rocks can transform from one type into another, as described by the geological model called the rock cycle. These events produce three general classes of rock: igneous, sedimentary, and **meta-morphic**.

C Igneous rock forms through the cooling and solidification of **magma** or **lava**. This magma can be derived from partial melts of pre-existing rocks in either a planet's mantle or crust. Typically, the melting of rocks is caused by one or more of three processes: an increase in temperature, a decrease in pressure, or a change in composition.

D Igneous rocks are divided into two main categories: **plutonic** rock and volcanic. Plutonic or **intrusive** rocks result when magma cools and crystallises slowly within Earth's crust. A common example of this type is granite. Volcanic or **extrusive** rocks result from magma reaching the surface either as lava or fragmental ejecta, forming minerals such as **pumice** or **basalt**. The chemical abundance and the rate of cooling of magma typically form a sequence known as Bowen's reaction series. Most major igneous rocks are found along this scale.

E About 64.7% of Earth's crust by volume consists of igneous rocks, making it the most plentiful category. Of these, 66% are basalts and **gabbros**, 16% are granite, and 17% granodiorites and **diorites**. Only 0.6% are **syenites** and 0.3% **peridotites** and **dunites**. The oceanic crust is 99% basalt, which is an igneous rock of **mafic** composition. Granites and similar rocks, known as meta-granitoids, form much of the continental crust. Over 700 types of igneous rocks have been described, most of them having formed beneath the surface of Earth's crust. These have diverse properties, depending on their composition and the temperature and pressure conditions in which they were formed.

F Sedimentary rocks are formed at Earth's surface by the accumulation and cementation of fragments of earlier rocks, minerals, and organisms or as chemical precipitates and organic growths in water (sedimentation). This process causes **clastic** sediments (pieces of rock) or organic particles (**detritus**) to settle and accumulate, or for minerals to chemically precipitate (evaporite) from a solution. The particulate matter then undergoes compaction and cementation at moderate temperatures and pressures (diagenesis).

metamorphic *adj.* 变形的,变质的;改变结构的
magma *n.* 岩浆
lava *n.* 熔岩;火山岩
plutonic *adj.* 深成的,火成的(岩石)
intrusive *adj.* 侵入的;闯入的,打扰的
extrusive *adj.* 挤出的,喷出的,突出的
pumice *n.* 轻石;浮石
basalt *n.* 玄武岩
gabbro *n.* 辉长岩
diorite *n.* 闪长岩
syenite *n.* 正长岩
peridotite *n.* 橄榄岩
dunite *n.* 纯橄榄岩
mafic *adj.* 铁镁质的
clastic *adj.* 可分解的,碎屑状的
detritus *n.* 碎石;(侵蚀形成的)岩屑;风化物

Before being deposited, sediments are formed by weathering of earlier rocks by erosion in a source area and then transported to the place of deposition by water, wind, ice, mass movement or glaciers (agents of denudation). Mud rocks comprise 65% (mudstone, **shale** and **siltstone**); sandstones 20% to 25% and carbonate rocks 10% to 15% (limestone and dolostone). About 7.9% of the crust by volume is composed of sedimentary rocks, with 82% of those being shales, while the remainder consists of limestone (6%), sandstone and arkoses (12%). Sedimentary rocks often contain fossils. Sedimentary rocks form under the influence of gravity and typically are deposited in horizontal or near horizontal layers or strata and may be referred to as stratified rocks. A small fraction of sedimentary rocks deposited on steep slopes will show cross bedding where one layer stops abruptly along an interface where another layer eroded the first as it was laid atop the first.

shale   n. 页岩；泥板岩
siltstone   n. 粉砂岩
metamorphism   n. 变质；变形
protolith   n. 原岩
intrusion   n. 干扰,干涉

G Metamorphic rocks are formed by subjecting any rock type—sedimentary rock, igneous rock or another older metamorphic rock—to different temperature and pressure conditions than those in which the original rock was formed. This process is called **metamorphism**; meaning to "change in form". The result is a profound change in physical properties and chemistry of the stone. The original rock, known as the **protolith**, transforms into other mineral types or other forms of the same minerals, by recrystallization. The temperatures and pressures required for this process are always higher than those found at Earth's surface: temperatures greater than 150 to 200°C and pressures of 1500 bars. Metamorphic rocks compose 27.4% of the crust by volume.

H The three major classes of metamorphic rock are based upon the formation mechanism. An **intrusion** of magma that heats the surrounding rock causes contact metamorphism—a temperature-dominated transformation. Pressure metamorphism occurs when sediments are buried deep under the ground; pressure is dominant, and temperature plays a smaller role. This is termed burial metamorphism, and it can result in rocks such as jade. Where both heat and pressure play a role, the mechanism is termed regional metamorphism. This is typically found in mountain-building regions.

I Depending on the structure, metamorphic rocks are divided into two general categories. Those that possess a texture are referred to as **foliated**; the remainders are termed non-foliated. The name of the rock is then determined based on the types of minerals present. Schists are foliated rocks that are primarily composed of **lamellar** minerals such as micas. A **gneiss** has visible bands of dif-fering lightness, with a common example being the granite gneiss. Other varieties of foliated rock include **slates, phyllites**, and **mylonite**. Familiar examples of non-foliated metamorphic rocks include **marble** and **soapstone**. This branch contains quartzite—a metamorphosed form of sandstone—and **hornfels**.

foliated   *v.* 有叶的,覆有叶的
lamellar   *adj.* 薄片状的;薄层状的
gneiss   *n.* 片麻岩
slate   *n.* 石板;板岩,页岩
phyllite   *n.* 千枚岩,硬绿泥石
mylonite   *n.* 糜棱岩
marble   *n.* 大理石
soapstone   *n.* 皂石;滑石
hornfels   *n.* 角页岩

Questions 1—4

Passage B has nie paragraphs, A—I.
Choose the correct heading for Paragraphs B and D—F from the list of headings below.
Write the correct number i—viii in boxes 1—4.

| 1 | 2 | 3 | 4 |

List of Headings

i. Igneous rocks are divided into two main categories
ii. Rocks are geologically classified according to characteristics
iii. How igneous rocks are formed
iv. How sedimentary rocks are formed
v. Why igneous rocks are the most plentiful category
vi. Metamorphic rocks are classified according to formation mechanism
vii. Definition of rock
viii. Metamorphism

Example	Paragraph **A**	Answer	vii
1	Paragraph **B**	Answer	
Example	Paragraph **C**	Answer	iii
2	Paragraph **D**	Answer	
3	Paragraph **E**	Answer	
4	Paragraph **F**	Answer	
Example	Paragraph **G**	Answer	viii
Example	Paragraph **H**	Answer	vi

Questions 5—10

Do the following statements reflect the claims of the writer in the Passage?
In boxes 5—10, write

  Y(YES)   if the statement reflects the claims of the writer
  N(NO)   if the statement contradicts the claims of the writer
  NG(NOT GIVEN) if it is impossible to say what the writer thinks about this

5		6		7		8		9		10	

5. Earth's outer solid layer, the lithosphere, is made of mineral.
6. The types and abundance of minerals in a rock are determined by the manner in which the rock was formed.
7. About 64.7% of Earth's crust by volume consists of sedimentary rocks.
8. Burial metamorphism can result in rocks such as jade.
9. Metamorphic rocks compose 27.4% of the crust by volume.
10. Burial metamorphism is more typically found than regional metamorphism.

Questions 11—13

Complete the summary of Paragraph G with the list of words A—H below. Write the correct letter A—H in boxes 11—13.

11		12		13	

  Metamorphic rocks are formed by subjecting sedimentary rock, igneous rock or another older metamorphic rock to different temperature and ___11___ conditions than those in which the original rock was formed. This process is called ___12___ meaning to "change in form". The three major classes of metamorphic rock are based upon the ___13___ mechanism.

A. minerals	B. organism	C. metamorphism	D. recrystallization
E. metamorphic	F. pressure	G. physical	H. formation

Unit 5
Geomorphology and Geography

 Scan and read along the passage

Geomorphology is the scientific study of landforms and the processes that shape them. Geomorphologists seek to understand why landscapes look the way they do: to understand landform history and dynamics, and predict future changes through a combination of field observation, physical experiment, and numerical modeling. Geomorphology is practiced within geography, geology, geodesy, engineering geology, archaeology, and geotechnical engineering[①].

Passage A

Geological Process—Weathering and Erosion[②]

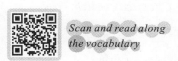 Scan and read along the vocabulary

A Rocks exposed on Earth's surface are constantly being altered by water, air, changing temperature, and other environmental factors. The term "weathering" refers to the group of destructive processes that change the physical and chemical character of rock on or near Earth's surface. The tightly bound crystals of an igneous rock[(1)] can be loosened and altered to new minerals by weathering. Weathering can be a mechanical or a chemical process. Often, these two types of weathering work together.

Mechanical weathering

B Mechanical or physical weathering involves the breakdown of rocks and soils through direct contact with atmospheric conditions

---

① This section is adapted from: Wikipedia. Geomorphology [EB/OL]. (2017-01-08)[2018-01-07]. https://en.wikipedia.org/wiki/Geomorphology.

② This passage is adapted from: 邹灿, 阎黎明. Text A Weathering and erosion[M]//新编地质英语教程. 高等学校专业英语教材. 北京: 电子工业出版社, 2013: 26-31.

such as heat, water, ice and pressure.

C Water seeps into cracks and **crevices** in rock. If the temperature drops low enough, the water will freeze. When water freezes, it expands. The ice then works as a **wedge**. It slowly widens the cracks and splits the rock. When ice melts, water performs the act of erosion by carrying away the tiny rock fragments lost in the split.[2]

D Mechanical weathering also occurs as rock heats up and cools down. The changes in temperature cause rock to expand and contact. As this happens over and over again, the rock weakens. Over time, it **crumbles**.

E Another type of mechanical weathering occurs when clay or other materials near hard rock absorb water. The clay swells with the water, breaking apart the surrounding rock[3].

F Salt also works to weather rock. Saltwater sometimes gets into the cracks and pores of rock. If the saltwater evaporates, salt crystals are left behind. As the crystals grow, they put pressure on the rock, slowly breaking it apart.

Chemical weathering

G The second classification, chemical weathering, involves the **decomposition** of rock from exposure to water and atmospheric gases (principally carbon dioxide and water vapor). As rock is **decomposed** by these agents, new chemical compounds form. Chemical weathering changes the materials that make up rocks and soil. Sometimes, carbon dioxide from air or soil combines with water. This produces a weak acid called carbonic acid[4], which can dissolve rock.

H Carbonic acid is especially effective at dissolving limestone. When the carbonic acid seeps through limestone underground, it can open up huge cracks or hollow out[5] vast networks of caves. Carlsbad Caverns National Park, in the State of New Mexico, USA, includes more than 110 limestone caves. The largest is called the Big Room. With about 1200m long and 190m wide, it is the size of six football fields.

I Sometimes, chemical weathering dissolves large regions of limestone or other rocks on the surface of Earth to form a landscape called **karst**. In these dramatic areas, the surface rock is

---

crevice *n.* 裂缝；裂隙
wedge *n.* 楔形体
crumble *vi.* 崩溃；破碎，崩解
  vt. 崩溃；弄碎，粉碎
decomposition *n.* 分解，腐败
decomposed *adj.* 已腐烂的，已分解的
karst *n.* 喀斯特；岩溶

pockmarked with holes, **sinkholes**, and eaves. One of the world's most spectacular examples of karst is Shilin, or the Stone Forest, near Kunming, China. Hundreds of slender, sharp towers of limestone rise from the landscape.

J Another type of chemical weathering works on rocks that contain iron. These rocks rust in a process called **oxidation**. As the rust expands, it weakens the rock and helps break it apart.

K Erosion is the process by which soil and rock are removed from Earth's surface by **exogenetic** processes such as wind or water flow, and then transported and deposited in other locations.

Erosion by water

L Moving water is the major agent of erosion. Rain carries away bits of soil and slowly washes away rock fragments. Rushing streams and rivers wear away[6] their banks, creating larger and larger valleys. In a span of about 5 million years, the Colorado River[7] cut deeper and deeper into the land in what is now the US state of Arizona. It eventually formed the Grand Canyon[8] which is more than 1600m deep and as much as 29km wide in some places.

M Erosion by water changes the shape of coastlines. Waves constantly crash against shores. They pound rocks into **pebbles** and **reduce** pebbles to sand. Water sometimes takes sand away from beaches. This moves the coastline farther inland.

N The battering of ocean waves also erodes seaside cliffs. It sometimes bores holes that form caves. When water breaks through the back of the cave, it creates an **arch**. The continual pounding of waves can cause the top of the arch to fall, leaving nothing but rock columns. These are called sea stacks[9]. All of these features make rocky beaches beautiful, but also dangerous.

Erosion by wind

O Wind is also an agent of erosion. It carries dust, sand, and volcanic ash from one place to another. Wind can sometimes blow sand into **towering dunes**. Some sand dunes in some areas of the Gobi Desert[10] in China reach more than 400m high.

P In dry areas, windblown sand blasts against rock with tremendous force, slowly wearing away the soft rock. It also polishes rocks and cliffs until they are smooth.

---

pockmark *n.* 麻子;凹坑
sinkhole *vt.* 使留下痘疤;使有凹坑
  n. 落水洞;灰岩坑
oxidation *n.* 〈化〉氧化
exogenetic *adj.* 外生的;外因的;外源性的
pebble *n.* 中砾,卵石
  v. (用卵石等)铺
reduce *vt.* 缩减;简化;还原
arch *n.* 背斜;穹隆;天生桥
  vt. 使……弯成弓形;用拱连接
  vi. 拱起;成为弓形
towering *adj.* 高耸的;卓越的;激烈的
dune *n.* (由风吹积而成的)沙丘

Q Wind is responsible for the dramatic arches that give Arches National Park, in the USA state of Utah, its name. Wind can also erode material until nothing remains at all. Over millions of years, wind and water eroded an entire mountain range in central Australia. Uluru(11), also known as Ayers Rock(12), is the only **remnant** of those mountains.

Erosion by ice

R Ice can erode land. In **frigid** areas and on some mountaintops, glaciers move slowly downhill and across the land. As they move, they pick up everything in their path, from tiny grains of sand to huge boulders.

S The rocks carried by a glacier rub against the ground below, eroding both the ground and the rocks. Glaciers grind up(13) rocks and scrape away the soil. Moving glaciers gouge out(14) basins and form steep—sided mountain valleys.

T Today, in places such as Greenland and Antarctica, glaciers continue to erode Earth. These ice sheets, sometimes more than a mile thick, carry rocks and other debris downhill toward the sea. Eroded sediment is often visible on and around glaciers. This material is called **moraine.**

U It is important to distinguish between weathering and erosion. Weathering breaks down rocks that are either stationary or moving. Erosion is the picking up or physical removal of rock particles by an agent such as streams, wind or glaciers. Weathering helps break down a solid rock into loose particles that are easily eroded. Most eroded rock particles are at least partially weathered, but rock can be eroded before it has been weathered at all. A stream can erode weathered or unweathered rock fragments. Far more erosion occurs naturally, and a combination of weathering and erosion is responsible for producing the soil from which Earth's plants grow.

V Weathering and erosion slowly **chisel**, polish, and **buff** Earth's rock into ever evolving works of art—and then wash the remains into the sea.

W Working together, they create and reveal marvels of nature from tumbling boulders high in the mountains to sandstone arches in the **parched** desert to polished cliffs braced against violent seas.

---

remnant   *n.* 剩余
adj. 剩余的
frigid   *adj.* 寒冷的, 严寒的
moraine   *n.* 冰碛, (熔岩流表面的) 火山碎屑
chisel   *vt.* 雕, 刻; 凿; 欺骗
n. 砾石滩
buff   *vt.* 有软皮摩擦; 缓冲; 擦亮, 抛光某物
n. 浅黄色; 软皮; 爱好者
parched   *adj.* 焦的; 炎热的; 炒过的; 干透的
vt. 烘干; 使极渴 (parch 的过去分词)

Notes

(1) igneous rock:火成岩。
(2) When ice melts, water performs the act of erosion by carrying away the tiny rock fragments lost in the split. 当冰融化,水便通过带走遗留在裂缝中的岩石小碎片履行其侵蚀作用。
(3) surrounding rock:围岩。
(4) carbonic acid:碳酸。
(5) hollow out:挖空。
(6) wear away:磨损;消磨;流逝。
(7) the Colorado River:(美国的)科罗拉多河。
(8) the Grand Canyon:(美国的)大峡谷。
(9) sea stacks:海蚀柱;海柱。
(10) the Gobi Desert:戈壁沙漠(蒙古和中国西北部)。
(11) Uluru:乌卢鲁(即澳洲艾尔斯岩,世界上最大的单体巨石)。
(12) Ayers Rock:(澳大利亚的)艾尔斯岩。
(13) grind up:磨碎。
(14) gouge out:挖出;凿槽。

Online Resources

1. Carlsbad Caverns National Park:Beauty and Wonder;Above and Below.(美国的)卡尔斯巴德洞窟国家公园。High ancient sea ledges, deep rocky canyons, flowering cactus and desert wildlife—treasures above the ground in the Chihuahuan Desert. Hidden beneath the surface are more than 119 caves, formed when sulfuric acid dissolved limestone leaving behind caverns of all sizes. Refer to https://www.nps.gov/cave/index.htm.

2. Arches National Park:A red rock wonderland.(美国的)拱门国家公园。Visit Arches and discover a landscape of contrasting colors, landforms and textures unlike any other in the world. The park has over 2000 natural stone arches, in addition to hundreds of soaring pinnacles, massive fins and giant balanced rocks. This red rock wonderland will amaze you with its formations, refresh you with its trails, and inspire you with its sunsets. Refer to https://www.nps.gov/arch/index.htm.

Exercises

Detailed Understanding

Ⅰ. Answer the following questions according to the passage you have read.
1. What is the definition of weathering?
2. How do atmospheric factors affect the process of mechanical weathering?
3. In what ways does chemical weathering work on rocks?
4. How many agents of erosion are mentioned here? What are they?
5. How does erosion by water change the shape of coastlines?
6. What forms sand dunes in some areas of the Gobi Desert?
7. How can ice erode the land?
8. How to distinguish between weathering and erosion?

Vocabulary

Ⅱ. Fill in the blanks with the words given below. Change the form where necessary. (15 个单词, 10 个空)

towering	deterioration	incorporate	decomposition	break down
fragment	frigid	oxidation	crumble	wear away
reduce	alter	parched	grind up	take place

1. The processes by which environmental agents at or near Earth's surface cause rocks and minerals to _____ is called weathering.
2. In _____ areas and on some mountaintops, glaciers move slowly downhill and across the land.
3. Countless monuments—from the pyramids of Egypt to ordinary tombstones—have suffered drastic _____ from freezing water, hot sunshine, and other climatic forces.
4. The waves are _____ the rocks.
5. Mechanical and chemical weathering _____ constantly and simultaneously in most environments.
6. This National Natural Landmark is known for its _____ sandstone and limestone formations, some of which reach heights of several hundred feet.
7. It frees life-sustaining minerals and elements from solid rock, allowing them to become _____ into our soils and finally into our foods.
8. The cumulative activities of even very small creatures, such as earthworms and insects, can also help break rocks into smaller _____.

9. Chemical weathering involves the _____ of rock from exposure to water and atmospheric gases (principally carbon dioxide and water vapor).
10. Chemical weathering _____ the composition of minerals and rocks principally through reactions involving water.

Sentence Structure

Ⅲ. Combine each pair of the following sentences, using an "as" clause to introduce action happening at the same time as another action.

Model

They move.

At the same time they pick up everything in their path, from tiny grains of sand to huge boulders.

⟶As they move, they pick up everything in their path, from tiny grains of sand to huge boulders.

---

1. Kate saw her brother Bill.
   At the same time she was getting off the school bus.

2. Mother was standing up from her seat.
   At the same time she dropped her glass to the ground.

3. A policeman fired at a thief.
   The thief was bolting out of the house at the time.

4. Sue ran to catch the school bus.
   At the same time she thought of her talk with her mother.

5. The teacher entered the classroom.
   All the students shouted "Happy Birthday" to him.

Translation
Ⅳ. Translate the following sentences into Chinese.
1. Weathering is a slow but potent force to which even the hardest rocks are susceptible.

2. Rocks that have been weakened by weathering are more vulnerable to erosion, the process by which gravity, moving water, wind, or ice transports pieces of rock and deposits them elsewhere.

3. Each of the different minerals within a rock expands to a different degree when heated, for example, quartz grains expand about three time as much as grains of plagioclase feldspar that are exposed to the seam heat.

4. Almost all rocks contain some cracks and crevices. Plants and trees may take root in such cracks in surface rocks.

5. Water is the single most important factor controlling the rate of chemical weathering, because it carries ions to the reaction site, participates in the reaction, and then carries away the products of the reaction.

V. Translate the following passage into English.

风化作用在我们的日常生活中扮演着至关重要的角色,既有积极影响又有消极后果。风化作用从固体岩石中释放了生命所必需的矿物和元素,使这些物质混合进入到土壤中,最终进入到我们的食物中。事实上,没有风化作用我们几乎得不到食物,正由于这一过程才产生了我们许多食物生长的各种土壤。但风化作用也对建筑物造成毁坏,数不清的纪念碑——从埃及的金字塔到平常的墓碑——都遭受了冰水、灼热的阳光以及其他气候营力的剧烈破坏。

Reading Skills

Recognising Differences Between Facts and Opinions[①]

Most passages contain ideas based on facts and opinions. It is very important to know when we are reading facts and when we are reading a writer's opinions. The ability to recognise differences between facts and opinions can help us to achieve a deeper understanding in our reading.

A *fact* is information that can be proved true through objective evidence. This evidence may be physical proof or the spoken or written testimony of witnesses.

The following are some more facts—they can be checked for accuracy and thus proved true.

Fact: The Quad Tower is the tallest building in the city.
(A researcher could go out and, through inspection, confirm that the building is the tallest.)

Fact: Einstein willed his violin to his grandson.
(This statement can be checked in historical publications or with Einstein's estate.)

Fact: On September 11, 2001, terrorists destroyed the New York World Trade Centre, killing thousands.
(This event was witnessed in person or on television by millions, and it is firmly in records worldwide.)

An *opinion* is a belief, judgment, or conclusion that cannot be objectively proved true. As a result, it is open to question. Remember that value (or judgment) often represent opinion. Here are examples of *value words*:

[①] This section is selectively adapted from: Langan J. Fact and opinion 事实和观点[M]//兰翠竹. 大学英语阅读进阶:英语技能提高丛书. 4版. 译注. 北京:外语教学与研究出版社, 2016:259-261.

| best | worse | better | worse | beautiful |
| great | terrible | lovely | disgusting | wonderful |

Here are some more opinions:

Opinion: The Quad Tower is the ugliest building in the city.

(There's no way to prove this statement because two people can look at the same building and come to different conclusions about its beauty. Ugly is a value word, a word we use to express a value judgment. Value or judgment words are signals that an opinion is being expressed. By their very nature, these words represent opinions, not facts.)

Opinion: Einstein should have willed his violin to a museum.

(Who says? Not his grandson. This is an opinion.)

Opinion: The attack on the World Trade Centre was the worst act of terrorism in the history of humankind.

(Whether something is "worst" is always debatable. Worst is another value word.)

Passage B

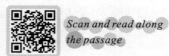
Scan and read along the passage

Disappearing Delta[①]

Scan and read along the vocabulary

A The **fertile** land of the Nile **delta** is being **eroded** along Egypt's **Mediterranean** coast at an **astounding** rate, in some parts estimated at 100m per year. In the past, land **scoured** away from the coastline by the currents of the Mediterranean Sea used to be replaced by sediment brought down to the delta by the River Nile, but this is no longer happening.

B Up to now, people have blamed this loss of delta land on the two large dams at Aswan in the south of Egypt, which hold back **virtually** all of the sediment that used to flow down the river. Before the dams were built, the Nile flowed freely, carrying huge quantities of sediment north from Africa's **interior** to be deposited on the Nile delta. This continued for 7000 years, eventually covering a region of over

fertile *adj.* 肥沃的
erode *v.* 腐蚀,水土流失
delta *n.* 三角洲
Mediterranean *adj.* 地中海的
astound *v.* 使惊讶
scour *v.* 冲刷
virtually *adv.* 差不多
interior *n.* 内陆
slit *n.* 淤泥

---

① This passage is selectively adapted from: University of Cambridge ESOL Examinations. Disappearing delta[M]//Cambridge IELTS 5 with answers. Cambridge: Cambridge University Press, 2006: 67 – 70.

22000 km² with layers of fertile **silt**. Annual flooding brought in new, **nutrient**-rich soil to the delta region, replacing what had been **washed away** by the sea, and dispensing with the need for **fertilisers** in Egypt's richest food-growing area. But when the Aswan dams were constructed in the 20th century to provide electricity and **irrigation**, and to protect the huge population centre of Cairo and its surrounding areas from annual flooding and drought, most of the sediment with its natural fertiliser accumulated up above the dam in the southern, upstream half of Lake Nasser, instead of passing down to the delta.

C Now, however there turns out to be more to the story. It appears that the sediment-free water emerging from the Aswan dams picks up silt and sand as it erodes the river bed and banks on the 800km trip to Cairo. Daniel Jean Stanley of the Smithsonian Institute noticed that water samples taken in Cairo, just before the river enters the delta, **indicated** that the river sometimes carries more than 850g of sediment per cubic metre of water—almost half of what it carried before the dams were built. "I'm ashamed to say that the significance of this didn't strike me until after I had read 50 or 60 studies," says Stanley in *Marine Geology*, "There is still a lot of sediment coming into the delta, but virtually no sediment comes out into the Mediterranean to **replenish** the coastline. So the sediment must be trapped on the delta itself."

D Once north of Cairo, most of the Nile water is **diverted** into more than 10 000km of irrigation **canals** and only a small proportion reaches the sea directly through the rivers in the delta. The water in the irrigation canals is still or very slow-moving and thus cannot carry sediment, Stanley explains. The sediment sinks to the bottom of the canals and then is added to fields by farmers or pumped with the water into the four large freshwater **lagoons** that are located near the outer edges of the delta. So very little of it actually reaches the coastline to replace what is being washed away by the Mediterranean currents.

E The farms on the delta plains and fishing and **aquacul**

nutrient　　*n.* 养料
fertiliser　　*n.* 肥料
irrigation　　*n.* 灌溉
indicate　　*v.* 表明
replenish　　*v.* 补充
divert　　*v.* 转移
canal　　*n.* 水道
lagoon　　*n.* 潟湖
aquaculture　　*n.* 水产养殖业

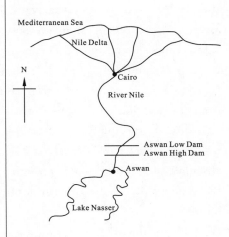

ture in the lagoons account for much of Egypt's food supply. But by the time the sediment has come to rest in the fields and lagoons it is laden with municipal, industrial and agricultural waste from the Cairo region, which is home to more than 40 million people. "Pollutants are building up faster and faster." says Stanley.

F Based on his **investigations** of sediment from the delta lagoons, Frederic Siegel of George Washington University concurs. "In Manzalah Lagoon, for example, the increase in **mercury**, **lead**, **copper** and **zinc** coincided with the building of the High Dam at Aswan, the availability of cheap electricity, and the development of major power-based industries," he says. Since that time the concentration of mercury has increased significantly. Lead from engines that use leaded fuels and from other industrial sources has also increased dramatically. These poisons can easily enter the food chain, affecting the productivity of fishing and farming. Another problem is that agricultural wastes include fertilisers which **stimulate** increases in plant growth in the lagoons and upset the ecology of the area, with serious effects on the fishing industry.

G According to Siegel, international environmental organizations are beginning to pay closer attention to the region, partly because of the problems of erosion and pollution of the Nile delta, but principally because they fear the impact this situation could have on the whole Mediterranean coastal ecosystem. But there are no easy solutions. In the immediate future, Stanley believes that one solution would be to make **artificial** floods to flush out the delta **waterways**, in the same way that natural floods did before the construction of the dams. He says, however, that in the long term an alternative process such as **desalination** may have to be used to increase the amount of water available. "In my view, Egypt must **devise** a way to have more water running through the river and the delta," says Stanley. Easier said than done in a desert region with a rapidly growing population.

---

investigation n. 调查
concur v. 同意
mercury n. 汞
lead n. 铅
copper n. 铜
zinc n. 锌
stimulate v. 刺激
artificial adj. 人造的
waterway n. 水道
desalination n. 海水淡化
devise n. 设计

Questions 1—5

Passage B has seven paragraphs, A—G.

Choose the correct heading for Paragraphs B and D—G from the list of headings below.

Write the correct number i—vii in boxes 1—5.

1		2		3		4		5	

List of Headings

i. Effects of irrigation on sedimentation
ii. Causing pollution in the Mediterranean
iii. Interrupting a natural processes
iv. The threat to food production
v. Less valuable sediment than before
vi. Egypt's disappearing coastline
vii. Looking at the long-term impact

Example	Paragraph A	Answer	vi
1	Paragraph B	Answer	
Example	Paragraph C	Answer	v
2	Paragraph D	Answer	
3	Paragraph E	Answer	
4	Paragraph F	Answer	
5	Paragraph G	Answer	

Questions 6—11

Do the following statements reflect the claims of writer in Passage B?

In boxes 6—11, write

    Y(YES)   if the statement reflects the claims of the writer
    N(NO)   if the statement contradicts the claims of the writer
    NG(NOT GIVEN) if it is impossible to say what the writer thinks about this

6		7		8		9		10		11	

6. Coastal erosion occurred along Egypt's Mediterranean coast before the building of the Aswan dams.

7. Some people predicted that the Aswan dams would cause land loss before they were built.

8. The Aswan dams were built to increase the fertility of the Nile delta.

9. Stanley found that the levels of sediment in the river water in Cairo were relatively high.
10. Sediment in the irrigation canals on the Nile delta causes flooding.
11. Water is pumped from the irrigation canals into the lagoons.

Questions 12－14

Complete the summary of Paragraphs E and F with the list of words A－H below.
Write the correct letter A－H in boxes 12－14.

| 12 | | 13 | | 14 | |

In addition to the problem of coastal erosion, there has been a marked increase in the level of __12__ contained in the silt deposited in the Nile delta To deal with this, Stanley suggests the use of __13__ in the short term, and increasing the amount of water available through __14__ in the longer term.

| A. artificial floods | B. desalination | C. delta waterways | D. natural floods |
| E. nutrients | F. pollutants | G. population control | H. sediment |

Unit 6
Climate Change and Atmosphere

Scan and read along the passage

People in the world are experiencing the change of climate: the chilly winter, the torrid summer, the sandstorm and the haze. Most people know it has something to do with industrial pollution and car exhaust. Global warming is one measure of climate change, which is a rise in the average global temperature.

Passage A

Effect of Climate Change[①]

Scan and read along the vocabulary

A The phrase climate change is used to describe a change in the climate, measured in terms of its **statistical properties**, e.g., the global mean surface temperature[(1)]. In this context, climate is taken to **mean** the average weather. Climate can change over period of time ranging from months to thousands or millions of years. The classical time period is 30 years, as defined by the World Meteorological Organization. The climate change referred to may be due to natural causes, e.g., changes in Sun's output, or due to human activities, e.g., changing the composition of the atmosphere. Any human-**induced** changes in climate will occur against the background of natural climatic **variations**.

B Climate change reflects a change in the energy balance of the climate system, i.e., changes the relative balance between incoming solar radiation and outgoing **infrared** radiation from Earth.

statistical *adj.* 统计的；统计学的
property *n.* 性质，性能；财产；所有权
mean *n.* 平均值
adj. 平均的；卑鄙的；低劣的；吝啬的
vi. 用意
vt. 意味；想要；意欲
induce *vt.* 诱导；引起；引诱；感应

[①] This passage is adapted from: Singh B R, Singh O. Study of impacts of global warming on climate change: rise in sea level and disaster frequency[M]//Global warming: impacts and future perspective. Rijeka, Croatia: InTech Publishing, 2012:94-118.

When this balance changes it is called "radiative forcing", and the calculation and measurement of radiative forcing is one aspect of the science of climatology[2]. The processes that cause such changes are called "forcing mechanisms"[3]. Forcing mechanisms can be either "**internal**" or "**external**". Internal forcing mechanisms are natural processes within the climate system itself, e.g., the **meridional** turnover. External forcing mechanisms can be either natural (e.g., changes in solar output) or anthropogenic (e.g., increased emissions of greenhouse gases).

C Whether the initial forcing mechanism is internal or external, the response of the climate system might be fast (e.g., a sudden cooling due to **airborne** volcanic ash reflecting sunlight), slow (e.g., **thermal** expansion of warming ocean water), or a combination (e.g., sudden loss of **albedo** in the **arctic** ocean as sea ice melts, followed by more gradual thermal expansion of the water). Therefore, the climate system can respond **abruptly**, but the full response to forcing mechanisms might not be fully developed for centuries or even longer.

D The most general definition of climate change is a change in the statistical properties of the climate system when considered over long periods of time, regardless of cause, whereas Global warming" refers to the change in Earth's global average surface temperature. Measurements show a global temperature increase of 1.4°F (0.78°C) between the years 1900 and 2005. Global warming is closely associated with a broad **spectrum** of other climate changes, such as:

- increases in the frequency of intense rainfall;
- decreases in snow cover and sea ice;
- more frequent and intense heat waves;
- rising sea levels;
- widespread ocean **acidification**.

Risk of intense rainfall

E There are two studied made here to **elaborate** the risk of intense rain fall one by United States and the other one by United Kingdom. They have warned that these risks are due to extreme climate change, thus we have to **curb** the global warming issues in phases. The summaries of study are given below:

---

variation   n. 变化；〈生〉变异,变种
infrared   n. 红外线
adj. 红外线的
internal   n. 内脏
adj. 内部的；里面的
external   adj. 外部的
n. 外部；外观；外面
meridional   adj. 南欧的；子午线的；南部的
airborne   adj. 〈航〉空运的；空气传播的；
thermal   adj. 热的；热量的；保热的
n. 上升的热气流
albedo   n. (行星等的)反射率；星体反照率(复数：albedos)
arctic   adj. 北极的；极寒的
n. 北极圈；御寒防水套鞋
abruptly   adv. 突然地；唐突地
spectrum   n. 光谱；频谱；范围；余象(复数：spectrums, spectra)
acidification   n. 〈化〉酸化；成酸性
elaborate   adj. 精心制作的；详尽的
vt. 精心制作；从简单成分合成(复杂有机物)
vi. 详细描述；变复杂
curb   n. 抑制；路边；勒马绳
vt. 控制；勒住

F Two 500-Year Floods in Just 15 Years: In the United States, The Great Flood of 1993— **devastating** communities along the Mississippi River and its **tributaries** in nine Midwestern states—was one of the most costly disasters. Thousands of Americans were displaced from their homes and forced to leave their lives behind, hundreds of **levees** failed, and damages soared to an estimated $12 billion to 16 billion. A mere 15 years later, history is repeating itself in the Midwest as the Rain Swollen Cedar, Illinois, Missouri and Mississippi Rivers and their tributaries top their banks and levees, leaving hundreds of thousands of people displaced. With rainfall in May—June 2008 about two to three times greater than the long-term average, soybean planting is behind schedule and some crops may have to be replanted. This remarkably quick return of such severe flooding is not **anticipated** by currently used out-of-date methodologies, but is what we should expect as global warming leads to more frequent and intense severe storms. Inadequate flood plain management is also responsible for the extent of damages from both floods, especially over-reliance on levees and the false sense of security they provide to those who live behind them. About 28% of the total new development in the seven states over the past 15 years has been in areas within the 1993 flood events.

G The National Wildlife Federation says that to limit the magnitude of changes to the climate and the impacts on communities and wildlife, we must curb global warming pollution. The National Wildlife Federation recommends that policy makers, industry, and individuals take steps to reduce global warming pollution from today's levels by 80% by 2050. That's a reduction of 20% per decade or just 2% per year. Science tells us that this is the only way to hold warming in the next century to no more than 2°F. This target is achievable with technologies either available or under development, but we need to start taking action now to avoid the worst impacts.

devastate　vt. 毁灭；毁坏；使荒废
tributary　adj. 纳贡的；附属的；辅助的
　　　　　n. 支流；进贡国；附属国
levee　n. 堤坝（码头）；（旧时君主或显贵的）早里接见
　　　vt. 为……筑堤
anticipate　vt. 预期，期望；占先，抢先；提前使用

Notes

(1) the global mean surface temperature：平均地表气温。

(2) climatology：This modern field of study is regarded as a branch of the atmospheric sciences

and a subfield of physical geography, which is one of Earth sciences. Climatology now includes aspects of oceanography and biogeochemistry. Basic knowledge of climate can be used within shorter term weather forecasting using analog techniques such as the El Niño—Southern Oscillation (ENSO), the Madden—Julian Oscillation (MJO), the North Atlantic Oscillation (NAO), the Northern Annular Mode (NAM) which is also known as the Arctic Oscillation (AO), the Northern Pacific (NP) Index, the Pacific Decadal Oscillation (PDO), and the Interdecadal Pacific Oscillation (IPO). Climate models are used for a variety of purposes from study of the dynamics of the weather and climate system to projections of future climate. Weather is known as the condition of the atmosphere over a period of time, while climate has to do with the atmospheric condition over an extended to indefinite period of time.

(3) forcing mechanisms：动力机制；驱动机制。

Online Resources

1. To know for more information about this text, please visit https://www.intechopen.com/books/global-warming-impacts-and-future-perspective/study-of-impacts-of-global-warming-on-climate-change-rise-in-sea-level-and-disaster-frequency.
2. Global Warming is caused by climate change. It is a serious problem that human beings are facing in 21st century. More information please go to http://www.nwf.org/globalwarming.

Exercises

Detailed Understanding

Ⅰ. Answer the following questions according to the passage you have read.
1. What does "climate" mean?
2. What is the classical time period of climate change?
3. What cause climate change?
4. What does climate change reflect?
5. What is radiative forcing?
6. How many types of forcing mechanism?
7. What does global warming mean?
8. What is global warming related with?
9. What is the target to reduce global warming pollution?

Unit 6 Climate Change and Atmosphere

Vocabulary

II. Fill in the blanks with the words given below. Change the form where necessary.（15 个单词，10 个空）

statistical	property	infrared	curb	anticipate
external	induce	spectrum	devastate	target
tributary	variation	elaborate	levee	internal

1. Decades ago, Kodak _____ that digital photography would overtake film.
2. Like fireplaces, they _____ a sense of comfort and warmth.
3. The survey found a wide _____ in the prices charged for canteen food.
4. The Missouri River is the chief _____ of the Mississippi.
5. Provide learners with opportunities to explain and _____ on what they're doing.
6. Isabel rose so _____ that she knocked down her chair.
7. We now have a Europe without _____ borders.
8. They intended to _____ the town at one stroke.
9. The term "special needs" covers a wide _____ of problems.
10. Mr. Obama's plan to _____ carbon dioxide (CO_2) emissions（排放）, though necessary, will be far from cost-free.

Sentence Structure

III. Combine each pair of the following sentences, using a "regardless" clause.

> **Model**
> Every problem you have is your responsibility.
> Who caused it.
> ⎯⎯→Every problem you have is your responsibility, regardless who caused it.

1. Who your stakeholders are.
   Get out there with them and see first-hand how they benefit from and/or struggle with your software.

2. Since all cloud systems perform under the same general concepts, this technique should be useful to you.
   Which cloud platform (or platforms) you choose to employ.

3. We have a unanimous vote for that candidate.
   How they finished in the Iowa caucuses.

4. To accurately reproduce the narrative in the output document, the style sheet must handle elements.
   Where they appear, and the push model excels at that.

5. Whether crime or misfortune was predominant in any given story.
   The faces in that world were specific and personal.

Ⅳ. **Translate the following sentences into Chinese.**
1. The phrase climate change is used to describe a change in the climate, measured in terms of its statistical properties, e.g., the global mean surface temperature.

2. The climate change referred to may be due to natural causes, e.g., changes in Sun's output, or due to human activities, e.g., changing the composition of the atmosphere.

3. Climate change reflects a change in the energy balance of the climate system, i.e., changes the relative balance between incoming solar radiation and outgoing infrared radiation from Earth.

4. The most general definition of climate change is a change in the statistical properties of the climate system when considered over long periods of time, regardless of cause, whereas "global warming" refers to the change in Earth's global average surface temperature.

5. The National Wildlife Federation says that to limit the magnitude of changes to the climate and the impacts on communities and wildlife, we must curb global warming pollution.

Ⅴ. **Translate the following passage into English.**
现在我们设想地球在复杂的物理、化学和生物过程的影响下是动态的、不断变化的、不断发展的,这些过程表明了速率和强度的很大变化。这些过程互相竞争,产生了趋向于一些平衡的情形或者是它们中的均衡。但是这种平衡总是受到新变化的干扰。这种变化的结果,部分是不能反复的,部分是可以反复的。因此,美国地质学家 Charles Richard van Hise 于 1898 年发表他的观点:"地球没有完成,现在正在重建,并且将来也永远要重建。"

Reading Skills

Guessing Meaning from Context

Guessing meaning from context refers to the ability to infer the meaning of an expression using contextual clues.

Techniques for guessing

Texts are often full of redundancy and consequently students can use the relation between different items within a text to get the meaning. Our prior knowledge of the world may also contribute to understanding what an expression means.

1. Synonyms and definitions
    • *Kingfishers* are a group of small to medium-sized brightly colored birds.
    • When he made *insolent* remarks towards his teacher, they sent him to the principal for being disrespectful.

2. Antonym and contrast

He *loved* her so much for being so kind to him. By contrast, he abhorred his mother.

3. Cause and effect

He was disrespectful towards other members. That's why he was sent off and *penalised*.

4. Parts of speech

Whether the word is a *noun*, a *verb*, an *adjective* or an *adverb*, it functions as a subject, a predicate or a complement.

5. Examples

Trojan is an example of computer virus.

6. Word forms (the morphological properties of the word)

Getting information from affixes (prefixes and suffixes) to understand a word. Examples: *dis-*(meaning not), *-less*(meaning without).

7. General knowledge

The French constitution establishes *laïcité* as a system of government where there is a strict *separation of church and state*.

These techniques help students get the meaning of words or at least narrow the possibilities. If need be, using the dictionary should be the last resort to fine tune the understanding of a vocabulary item.

Passage B

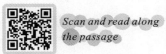

Scan and read along the passage

Sun's Fickle Heart May Leave Us Cold[①]

Stuart Clark

Scan and read along the vocabulary

A There's a **dimmer** switch inside Sun that causes its brightness to rise and fall on timescales of around 100 000 years -exactly the same period as between ice ages on Earth. So says a physicist who has created a computer model of our star's core.

B Robert Ehrlich of George Mason University in Fairfax, Virginia, modelled the effect of temperature fluctuations in Sun's interior. According to the standard view, the temperature of Sun's core is held constant by the opposing pressures of gravity and nuclear fusion. However, Ehrlich believed that slight variations should be possible.

C He took as his starting point the work of Attila Grandpierre

dimmer n. (车辆)调光器;光暗掣;衰减器

[①] This passage is adapted from: Clark S. Sun's fickle heart may leave us cold[N]. New Scientist, 2017-01-25.

of the Konkoly Observatory of the Hungarian Academy of Sciences. In 2005, Grandpierre and a **collaborator**, Gábor Ágoston, calculated that **magnetic** fields in Sun's core could produce small instabilities in the solar **plasma**. These instabilities would induce localised oscillations in temperature.

D Ehrlich's model shows that whilst most of these **oscillations** cancel each other out, some reinforce one another and become long-lived temperature variations. The favoured frequencies allow Sun's core temperature to oscillate around its average temperature of 13.6 million kelvin in cycles lasting either 100 000 or 41 000 years. Ehrlich says that random interactions within Sun's magnetic field could **flip** the **fluctuations** from one cycle length to the other.

E These two timescales are instantly recognizable to anyone familiar with Earth's ice ages: for the past million years, ice ages have occurred roughly every 100 000 years. Before that, they occurred roughly every 41 000 years.

F Most scientists believe that the ice ages are the result of subtle changes in Earth's orbit, known as the Milankovitch cycles. One such cycle describes the way Earth's orbit gradually changes shape from a circle to a slight **ellipse** and back again roughly every 100 000 years. The theory says this alters the amount of solar radiation that Earth receives, triggering the ice ages. However, a persistent problem with this theory has been its inability to explain why the ice ages changed frequency a million years ago.

G "In Milankovitch, there is certainly no good idea why the frequency should change from one to another," says Neil Edwards, a climatologist at the Open University in Milton Keynes, UK. Nor is the transition problem the only one the Milankovitch theory faces. Ehrlich and other critics claim that the temperature variations caused by Milankovitch cycles are simply not big enough to drive ice ages.

H However, Edwards believes the small changes in solar heating produced by Milankovitch cycles are then **amplified** by feedback mechanisms on Earth. For example, if sea ice begins to form because of a slight cooling, carbon dioxide that would otherwise have found its way into the atmosphere as part of the carbon cycle is locked into the ice. That weakens the greenhouse effect and Earth grows even colder.

collaborator　*n.*〈劳经〉合作者；勾结者；通敌者
magnetic　*adj.*地磁的；有磁性的；有吸引力的
plasma　*n.*〈等离子〉等离子体；血浆；〈矿物〉深绿玉髓
oscillation　*n.*振动；波动；动摇；〈物〉振荡
flip　*vt.*轻击；急挥
　　　*vi.*发疯；捻；蹦蹦跳跳
　　　*n.*浏览；空翻；跳跃
　　　*adj.*无礼的；冒失的
ellipse　*n.*〈数〉椭圆形，〈数〉椭圆
amplify　*v.*放大；详述
　　　*adj.*放大的；扩充的

I According to Edwards, there is no lack of such mechanisms. "If you add their effects together, there is more than enough feedback to make Milankovitch work," he says, "The problem now is identifying which mechanisms are at work." This is why scientists like Edwards are not yet ready to give up on the current theory. "Milankovitch cycles give us ice ages roughly when we observe them to happen. We can calculate where we are in the cycle and compare it with observation," he says. "I can't see any way of testing Ehrlich's idea to see where we are in the temperature oscillation."

J Ehrlich concedes this. "If there is a way to test this theory on Sun, I can't think of one that is practical," he says. That's because variation over 41 000 to 100 000 years is too gradual to be observed. However, there may be a way to test it in other stars: red **dwarfs**. Their cores are much smaller than that of Sun, and so Ehrlich believes that the oscillation periods could be short enough to be observed. He has yet to calculate the precise period or the extent of variation in brightness to be expected.

K Nigel Weiss, a solar physicist at the University of Cambridge, is far from convinced. He describes Ehrlich's claims as "**utterly implausible**". Ehrlich counters that Weiss' opinion is based on the standard solar model, which fails to take into account the magnetic instabilities that cause the temperature fluctuations.

dwarf *vi.* 变矮小
n. 侏儒，矮子
vt. 使矮小
adj. 矮小的
utterly *adv.* 完全地；绝对地；全然地；彻底地，十足地
implausible *adj.* 难以置信的，不像真实的

Questions 1–4

Complete each of the following statements with One or Two names of the scientists from the box below.

A. Attila Grandpierre	B. Gábor Ágoston	C. Neil Edwards
D. Nigel Weiss	E. Robert Ehrlich	

1. _____ claims there is a dimmer switch inside Sun that causes its brightness to rise and fall in periods as long as those between ice ages on Earth.
2. _____ calculated that the internal solar magnetic fields could produce instabilities in the solar plasma.
3. _____ holds that Milankovitch cycles can induce changes in solar heating on Earth and the changes are amplified on Earth.
4. _____ doesn't believe in Ehrlich's viewpoints at all.

Unit 6　Climate Change and Atmosphere

Questions 5—9

Do the following statements agree with the information given in the passage?

　　T(TRUE)　　　　if the statement is true according to the passage
　　F(FALSE)　　　if the statement is false according to the passage
　　NG(NOT GIVEN)　if the information is not given in the passage

5		6		7		8		9	

5. The ice ages changed frequency from 100 000 to 41 000 years a million years ago.

6. The sole problem that the Milankovitch theory can not solve is to explain why the ice age frequency should shift from one to another.

7. Carbon dioxide can be locked artificially into sea ice to eliminate the greenhouse effect.

8. Some scientists are not ready to give up the Milankovitch theory though they have not figured out which mechanisms amplify the changes in solar heating.

9. Both Edwards and Ehrlich believe that there is no practical way to test when the solar temperature oscillation begins and when ends.

Questions 10—14

Complete the notes below.

Choose one suitable word from the passage above for each answer.

10		11		12		13		14	

　　The standard view assumes that the opposing pressures of gravity and nuclear fusions hold the temperature ___10___ in Sun's interior, but the slight changes in Earth's ___11___ alter the temperature on Earth and cause ice ages every 100 000 years. A British scientist, however, challenges this view by claiming that the internal solar magnetic ___12___ can induce the temperature oscillations in Sun's interior. Sun's core temperature oscillates around its average temperature in ___13___ lasting either 100 000 or 41 000 years. And the ___14___ interactions within Sun's magnetic field could flip the fluctuations from one cycle length to the other, which explains why the ice ages changed frequency a million years ago.

Unit 7
Natural Resources and Environment Protection

 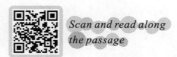 Scan and read along the passage

Oil includes classes of chemical compounds that may be otherwise unrelated in structure, properties, and uses. Oils may be animal, vegetable, or petrochemical in origin, and may be volatile or non-volatile. They are used for food (e.g., olive oil), fuel (e.g., heating oil), medical purposes (e.g., mineral oil), lubrication (e.g., motor oil), and the manufacture of many types of paints, plastics, and other materials.

Passage A

Oil and Gas[①]

 Scan and read along the vocabulary

hydrocarbon　n.〈化〉碳氢化合物
nitrogen　n.〈化〉氮
impurity　n. 杂质；不纯；不洁
flourish　n. 兴旺；茂盛
　vi. 繁荣，兴旺；茂盛；活跃；处于旺盛时期
eon　n. 永世；无数的年代；极长时期
bacterial　adj.〈微〉细菌的

A　Petroleum (rock-oil, from the Latin "petra"= rock or stone, and "oleum"= oil) is widespread throughout the world. It may be a gas, liquid, semi-solid, solid, or in more than one of these states at a single place. Any petroleum is a complex chemical mixture of **hydrocarbons**, which are compounds composed mainly of hydrogen and carbon, with smaller amounts of **nitrogen**, oxygen, and sulfur as **impurities**.

B　Scientists think that petroleum formation began many millions of years ago, when lower forms of plants and animals **flourished** in and near the oceans, as they do today. When these organisms died, their remains settled to the ocean bottoms where they gradually were deeply buried in mud and silt. Over **eons** of time, this abundant organic matter was transformed into oil and natural gas by high temperatures and pressures, decay, and **bacterial**

[①] This passage is adapted from: Lane E. Florida's geological history and geological resources[R]. Tallahassee, FL, USA: Florida Geological Survey, 1994.

processes, in a natural pressure cooker. At the same time, the enclosing sediments also were being transformed into consolidated rocks, such as sandstone, shale, or **limestone**. These rocks, in which the oil was formed, are called source rocks.

C Contrary to popular belief, oil does not occur in underground, **cistern**-like "pools" that can be tapped and pumped dry. Pool is a term that has special meaning in the oil industry; it refers to an economically producible quantity of oil **dispersed** in rock within Earth. Rock **strata** that contain economically recoverable concentrations of oil is called **reservoirs**.

D In order for oil to be concentrated in **porous** reservoir rocks, natural traps, **seals**, or cap rocks must occur, in various forms. In south Florida the oil traps are due to denser, less **permeable** rocks that overlie the oil fields' reservoir rocks. The traps in the north Florida **panhandle** fields are due to very impermeable beds of **anhydrite** (evaporitic salts), **faulting**, and **stratigraphic** traps.

E During the course of oil formation and accumulation in reservoirs, some of the original sea water was displaced and gravity separated the gas, oil, and water into layers, but in reality the situation within a reservoir is much more complex. Oil is only a small fraction of the fluids in the **pores** of a reservoir, but the discovery and recovery of this small fraction is the basis of the oil industry—and most of the world's energy. Most of the contained fluid is salt water, or **brine**, since its dissolved salt content may be higher than in sea water. Almost all crude oil has some gas dissolved in it under pressure. In some cases, excess gas forms a "gas-cap" above the oil zone. Some pores may contain only oil, or only gas, or only brine, or mixtures of all. Some of the oil is coated on the rock, while some is **suspended** in the brine. If a well were to **penetrate** this zone, the pressure would try to drive the oil, gas, and brine out of the rock and into the well. Not all of the gas and liquids would be driven out, however, no matter how great the driving pressure. Much of the oil would still remain in the rock due to **capillary** and **molecular** attraction between the rock and oil. Several techniques have been devised to increase the yield of oil from reservoirs, such as water, steam, or gas injection, and even **igniting** some of the oil, but recovery usually is relatively low; a recovery of 30% to 40% of the in-place oil is considered good.

limestone *n.*〈岩〉石灰岩
cistern *n.* 水箱;水池;贮水器
disperse *vt.* 分散;使散开
vi. 分散
adj. 分散的
reservoir *n.*〈水利〉水库;油箱;〈化〉储液器
porous *adj.* 多孔渗水的;能渗透的;有气孔的
seal *n.* 密封;印章
vt. 密封;盖章;海豹;封条;标志
permeable *adj.* 能透过的;有渗透性的
panhandle *n.* 平锅柄,〈美〉柄状的狭长区域
vt. 向……乞讨
vi. 行乞
anhydrite *n.*〈矿物〉硬石膏;〈建〉无水石膏
stratigraphic *adj.* 地层的;地层学的
pore *vi.* 细想;凝视
n. 气孔;小孔
vt. 使注视
brine *n.* 卤水;海水
vt. 用浓盐水处理
suspend *vt.* 延缓;推迟;使暂停
vi. 悬浮
penetrate *vt.* 渗透;穿透;洞察
vi. 渗透;刺入;看透
capillary *n.* 毛细管
adj. 毛细管的;毛状的
molecular *adj.*〈化〉分子的;由分子组成的
ignite *vt.* 点燃;使燃烧;使激动
vi. 点火

F There are two oil-producing areas in Florida. One is in south Florida, with 14 fields, and the other is in the western panhandle, with seven fields. The south Florida fields are located in Lee, Hendry, Collier, and Dade Counties. Florida's first oil field, the Sunniland field, in Collier County, was discovered in 1943. It has since produced over 18 million barrels of oil. Subsequently, 13 more field discoveries were found to lie along the northwest-southeast trend through Lee, Hendry, Collier, and Dade Counties. Although these fields are relatively small, production is significant. Together, the three Felda fields (West Felda, Mid-Felda, and Sunoco Felda) in Hendry County have produced over 54 million barrels of oil.

G The **depositional** environment during the Lower Cretaceous in south Florida was one of a shallow sea with a very slowly subsiding sea bottom. The time interval was characterised by numerous **transgressions** and regressions of the sea over the land, which created the carbonate-evaporite sequence[1] of geologic formations. The Sunniland "**reefs**" are not true patch reefs but were localised mounds of marine animals and **derris** on the sea floor. The primary **mound-builders** found in the Sunniland limestone were **rudistids**, **oyster-like mollusks** that existed only during the Cretaceous. They lived in great **profusion** and were widely distributed in clear, shallow **Cretaceous** seas. Other marine life found in the Sunniland patch reefs, or **mounds**, included **calcareous algae**, seaweed, **foraminifera**, and **gastropods**, such as snails.

depositional   *adj.* 沉积作用的
transgression   *n.* 〈地质〉海侵;犯罪;违反;逸出
reef   *n.* 暗礁,〈地质〉矿脉;收帆
derris   *n.* 〈植〉鱼藤;鱼藤属
mound-builder   *n.* 筑墩人
oyster-like   *n.* 软体动物
mollusk   *n.* 〈动〉软体动物
profusion   *n.* 丰富,充沛;慷慨
calcareous   *adj.* 钙质的,石灰质的   *n.* 钙化软骨
mound   *n.* 堆;高地;坟堆;护堤
calcareous algae   *n.* 钙性藻类,石灰藻,石灰质海草,钙藻类
algae   *n.* 〈植〉藻类;〈植〉海藻
foraminifera   *n.* 有孔虫
gastropod   *n.* 腹足类动物

Notes

(1) carbonate-evaporite sequence:碳酸盐岩-蒸发岩层序。碳酸盐岩层序是指碳酸盐沉积物由近滨岸的高能浅滩颗粒灰岩向海方向逐渐变成较深水碳酸盐沉积物的层序。随着层序地层研究的不断深入,关于碳酸盐岩岩溶储层的研究发现,无论是形成于向上变浅的米级旋回顶部的岩溶,还是由于局部性大地构造作用或海平面下降使得大部分碳酸盐岩台地暴露于大气淡水作用带下形成的局部性岩溶,这些现象说明,层序地层格架对碳酸盐岩岩溶储层的发育具有重要的控制作用。在碳酸盐岩中,4个主要变量控制着地层分布模式的变化和岩相分布:①构造沉降,它产生了沉积物的沉积空间;②全球海平面升降变化,它被认为是控制地层分布模式和岩相分布的主要控制因素;③沉积物的供应量,它控制古水深;④气候,它是控制沉积物类型的主要因素,其中降雨量和温度对碳酸盐岩、蒸发岩的分布起重要的控制作用。

Unit 7 Natural Resources and Environment Protection

Online Resources

To know more information about this text, please visit http://www.dep.state.fl.us/geology.

Exercises

Detailed Understanding

Ⅰ. Answer the following questions according to the passage you have read.

1. How long is the formation of oil?
2. What are "source rocks"? Please give some examples.
3. What does 'pool' refer to?
4. What are "reservoirs"?
5. What are the oil traps in the north Florida fields?
6. People could drive out all the oil with great enough driving pressure. Is it right?
7. To increase the yield of oil, what techniques are invented?
8. How much can people get from recovery?

Vocabulary

Ⅱ. Fill in the blanks with the words given below. Change the form where necessary. (15 个单词, 10 个空)

flourish	penetrate	disperse	seal	fault
mound	pore	suspend	penetrate	ignite
transgression	profusion	permeable	porous	sulfur

1. If a building, area, or country _____, no one can enter or leave it.
2. Cornflowers grow in _____ in the fields.
3. Somebody may say that sin is the _____ the law.
4. The army could still be used to _____ protesters.
5. Like I say, you have to _____ disbelief.
6. Because water is the softest substance in the world, but yet it can _____ the hardest rock or anything.
7. Plants will not _____ without water.
8. It can _____ your spirit and make you feel like you can do anything.
9. Not until we pointed out their _____ to them did they realise it.
10. Asian girl's skin tends to be very smooth and _____-less.

Sentence Structure

Ⅲ. Combine each pair of the following sentences, using a "contrary to" clause.

> **Model**
> Love (as with every other emotion we feel) is not rooted in the heart, but in the brain.
> The anatomy referenced in all of our favorite love songs.
> ⟶Contrary to the anatomy referenced in all of our favorite love songs, love (as with every other emotion we feel) is not rooted in the heart, but in the brain.

1. The common description of the cold war as a conflict between east and west.
    Communism was a prototypically western ideology, with roots in the European enlightenment.

2. Those people, many others think advertisements are very unpleasant.
    Consumers are often cheated by the false advertisement on which consumers always waste a great deal of time.

3. It also shows that agriculture in developed countries will not be decimated.
    Claims of vested farm interests in those countries.

4. What you may have seen recently.
    There's an upward bias to the stock market.

5. We can't be everywhere at once.
    To popular belief, you can not always do everything you want to do.

Unit 7　Natural Resources and Environment Protection

IV. Translate the following sentences into Chinese.

1. Scientists think that petroleum formation began many millions of years ago, when lower forms of plants and animals flourished in and near the oceans, as they do today.

2. Contrary to popular belief, oil does not occur in underground, cistern-like "pools" that can be tapped and pumped dry.

3. During the course of oil formation and accumulation in reservoirs, some of the original sea water was displaced and gravity separated the gas, oil, and water into layers, but in reality the situation within a reservoir is much more complex.

4. Not all of the gas and liquids would be driven out, however, no matter how great the driving pressure.

5. If a well were to penetrate this zone, the pressure would try to drive the oil, gas, and brine out of the rock and into the well.

V. Translate the following passage into English.

　　其中的一些工作是基于最新出品的工业数据集而展开研究的，而另一些则基于纯粹的学术渊源。加在一起使我们对于该地区的理解有显著的提高，同时也为即将到来的由国际海洋发现计划所提出的于2014年2月至3月在南海深水科学钻井奠定了基础。虽然对于在开放期间如何应变压力仍有争论，但是很明显，中国南海的生成是温暖的大陆地壳延伸的结果，这也是受早期阶段被称为华夏弧的华南板块向下俯冲的影响。

Reading Skills

Skimming and Scanning

Skimming and scanning are two very different strategies for speed reading. They are each used for different purposes, and they are not meant to be used all the time. They are at the fast end of the speed reading range, while studying is at the slow end.

Skimming refers to looking only for the general or main ideas, and works best with non-fiction (or factual) material. With skimming, your overall understanding is reduced because you don't read everything. You read only what is important to your purpose. Skimming takes place while reading and allows you to look for details in addition to the main ideas.

Unlike skimming, when scanning, you look only for a specific fact or piece of information without reading everything.

Passage B

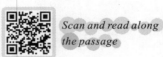
Scan and read along the passage

Undersea turbines which produce electricity from the tide's are set to become an important source of renewable energy for Britain. It is still too early to predict the extent of the impact they may have, but all the signs are that they will play a significant role in the future.

Tidal Power[①]

Scan and read along the vocabulary

Ⓐ Operating on the same principle as wind **turbines**, the power of sea turbines comes from **tidal** currents which turn blades similar to ships' **propellers**, but, unlike wind, the tides are predictable and the power input is constant. The technology raises the prospect of Britain becoming in renewable energy and **drastically** reducing its carbon dioxide emissions. If tide, wind and wave power are all

turbine n. 汽轮机;涡轮机;透平机
tidal   adj. 潮汐的;潮水的;由于潮水作用的;定时涨落的
propeller n. 螺旋桨,推进器
drastically adv. 大大地,彻底地;激烈地

---

① This passage is adapted from: University of Cambridge ESOL Examinations. Tidal power[M]//Cambridge IELTS 9 with answers. Cambridge:Cambridge University Press,2010:67-70.

developed, Britain would be able to close gas, coal and nuclear power plants and export renewable power to other parts of Europe. Unlike wind power, which Britain originally developed and then abandoned for 20 years allowing the Dutch to make it a major industry, undersea turbines could become a big export earner to island nations, such as Japan and New Zealand.

B Tidal sites have already been identified that will produce one sixth or more of the UK's power—and at prices competitive with modern gas turbines and **undercutting** those of the already **ailing** nuclear industry. One site alone, the Pentland Firth, between Orkney and mainland Scotland, could produce 10% of the country's electricity with banks of turbines under the sea, and another at Alderney in the Channel Islands three times the 1200 **megawatts** of Britain's largest and newest nuclear plant, Sizewell B, in Suffolk. Other sites identified include the Bristol Channel and the west coast of Scotland, particularly the channel between Campbelltown and Northern Ireland.

C Work on designs for the new turbine **blades** and sites are well advanced at the University of Southampton's sustainable energy research group. The first station is expected to be installed off Lynmouth in Devon shortly to test the technology in a venture **jointly** funded by the department of Trade and Industry and the European Union. AbuBakr Bahaj, in charge of the Southampton research, said: "The prospects for energy from tidal currents are far better than from wind because the flows of water are predictable and constant. The technology for dealing with the hostile **saline** environment under the sea has been developed in the North Sea oil industry and much is already known about turbine blade design, because of wind power and ship propellers. There are a few technical difficulties, but I believe in the next five to ten years we will be installing commercial marine turbine farms." Southampton has been awarded £215 000 over three years to develop the turbines and is working with Marine Current Turbines, a subsidiary of IT power, on the Lynmouth project. EU research has now identified 106 potential srfes for tidal power, 80% round the coasts of Britain. The best sites are between islands or around heavily indented coasts where there are strong tidal currents.

D A marine turbine blade needs to be only one third of the size

undercut　v.(网球等)从下削球；廉价出售；较便宜的工资工作
ailing　adj.生病的；不舒服的
megawatt　n.兆瓦,百万瓦特(电能计量单位)
blade　n.桨叶；刀片,剑；(壳、草等的)叶片
jointly　adv.共同地,联合地,连带地
saline　adj.含盐的,咸的　n.盐湖,盐泉；盐溶液

of a wind **generator** to produce three times as much power. The blades will be about 20m in **diameter**, so around 30m of watar is required. Unlike wind power, there are unlikely to be environmental objections. Fish and other creatures are thought unlikely to be at risk from the relatively slow-turning blades. Each turbine will be mounted on a tower which will connect to the national power supply **grid** via underwater **cables**. The towers will stick out of the water and be lit, to warn shipping, and also be designed to be lifted out of the water for maintenance and to clean **seaweed** from the blades.

E Dr. Bahaj has done most work on the Alderney site, where there are powerful currents. The single undersea turbine farm would produce far more power than needed for the Channel Islands and most would be fed into the French Grid and be reimported into Britain via the cable under the Channel.

F One technical difficulty is **cavitation**, where low pressure behind a turning blade causes air **bubbles**. These can cause vibration and damage the blades of the turbines. Dr. Bahaj said: "We have to test a number of blade types to avoid this happening or at least make sure it does not damage the turbines or reduce performance. Another slight concern is **submerged debris** floating into the blades. So far we do not know how much of a problem it might be. We will have to make the turbines **robust** because the sea is a hostile environment, but all the signs that we can do it are good".

Questions 1－4

Passage B has six paragraphs, A－F. Which paragraph contains the following information?
Write the correct letter, A－F, in boxes 1－4. You may use any letter more than once.

1		2		3		4	

1. The location of the first test site.
2. A way of bringing the power produced on one site back into Britain.
3. A reference to a previous attempt by Britain to find an alternative source of energy.
4. Mention of the possibility of applying technology from another industry.

Questions 5－9

Choose FIVE letters from the following, A－J. Write the correct letter in boxes 5－9.
Which FIVE of the following claims about tidal power are made by the writer?

5		6		7		8		9	

A. It is a more reliable source of energy than wind power.
B. It would replace all other forms of energy in Britain.
C. Its introduction has come as a result of public pressure.
D. It would cut down on air pollution.
E. It could contribute to the closure of many existing power stations in Britain.
F. It could be a means of increasing national income.
G. It could face a lot of resistance from other fuel industries.
H. It could be sold more cheaply than any other type of fuel.
I. It could compensate for the shortage of inland sites for energy production.
J. It is best produced in the vicinity of coastlines with particular features.

Questions 10－13

Label the diagram below.

Choose NO MORE THAN TWO words from the passage for each answer.

Write your answers in boxes 10－13.

| 10 | | 11 | | 12 | | 13 | |

Whole tower can be raised for __10__ and the extraction of seaweed from the blades.

Sea life not in danger due to the fact that blades are comparatively __11__.

Air bubbles result from the __12__ behind blades. This is known as __13__.

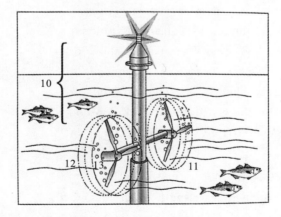

An undersea turbine

Unit 8
Geological Hazard

  Scan and read along the passage

A geological hazard is one of several types of adverse geologic conditions capable of causing damage or loss of property and life. These hazards include earthquakes, earthquake-triggered phenomena such as tsunamis, landslide, avalanches and its runout.

Passage A

Tsunami[①]

 Scan and read along the vocabulary

A A **tsunami** is a series of waves in a water body caused by the **displacement** of a large volume of water, generally in an ocean or a large lake. Earthquakes, volcanic eruptions, landslides, glacier **calvings**, and other disturbances above or below water all have the potential to generate a tsunami.[(1)] Unlike normal ocean waves which are generated by wind, or tides which are generated by the **gravitational** pull of Moon and Sun, a tsunami is generated by the displacement of water.

tsunami n. 海啸
displacement n. 取代，替代
calving n. 裂冰（作用）
v. （冰川）崩解
gravitational adj. 万有引力的，重力的
wavelength n. 波段；波长

B Tsunami waves do not resemble normal undersea currents or sea waves, because their **wavelength** is far longer. Rather than appearing as a breaking wave, a tsunami may instead initially resemble a rapidly rising tide, and for this reason they are often referred to as tidal waves, although this usage is not favoured by the scientific community because tsunamis are not tidal in nature. Tsunamis generally

---

①This passage is adapted from: Wikipedia. Tsunami [EB/OL]. (2017-12-29)[2017-12-31]. https://en.wikipedia.org/wiki/Tsunami.

consist of a series of waves with periods ranging from minutes to hours, arriving in a so-called "internal wave train". Wave heights of tens of metres can be generated by large events. Although the impact of tsunamis is limited to coastal areas, their destructive power can be enormous and they can affect entire ocean **basins**. The 2004 Indian Ocean tsunami was among the deadliest natural disasters in human history with at least 230 000 people killed or missing in 14 countries bordering the Indian Ocean.

C The term seismic sea wave also is used to refer to the phenomenon, because the waves most often are generated by seismic activity such as earthquakes. Prior to the rise of the use of the term tsunami in English-speaking countries, scientists generally encouraged the use of the term seismic sea wave rather than tidal wave. However, like tsunami, seismic sea wave is not a completely accurate term, as forces other than earthquakes—including underwater landslides, volcanic eruptions, underwater explosions, land or ice slumping into the ocean and the weather when the atmospheric pressure changes very rapidly—can generate such waves by displacing water.

D The principal generation mechanism or cause of a tsunami is the displacement of a substantial volume of water or **perturbation** of the sea. This displacement of water is usually attributed to either earthquakes, landslides, volcanic eruptions, glacier calvings or more rarely by meteorites and nuclear tests. The waves formed in this way are then sustained by gravity. Tides do not play any part in the generation of tsunamis.

E Tsunami can be generated when the sea floor abruptly deforms and vertically displaces the overlying water. Tectonic earthquakes are a particular kind of earthquake that are associated with Earth's crustal deformation; when these earthquakes occur beneath the sea, the water above the deformed area is displaced from its **equilibrium** position. More specifically, a tsunami can be generated when **thrust** faults associated with **convergent** or destructive plate boundaries move abruptly, resulting in water displacement, owing

basin n. 盆地;流域
perturbation n. 忧虑;不安;烦恼;摄动
equilibrium n. 平衡,均势;平静
thrust n. 推力;逆断层
convergent adj. 趋同的;会合的;逐渐减小的

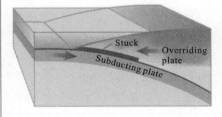

Drawing of tectonic plate boundary before earthquake

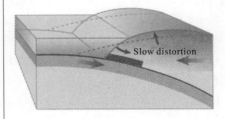

Overriding plate bulges under strain, causing tectonic uplift

to the vertical component of movement involved. Movement on normal (extensional) faults can also cause displacement of the seabed, but only the largest of such events (typically related to **flexure** in the outer trench swell) cause enough displacement to give rise to a significant tsunami, such as the 1933 Sanriku events.

F Tsunamis have a small **amplitude** (wave height) offshore, and a very long wavelength (often hundreds of kilometres long, whereas normal ocean waves have a wavelength of only 30m or 40m), which is why they generally pass unnoticed at sea, forming only a slight swell usually about 300mm above the normal sea surface. They grow in height when they reach shallower water, in a wave shoaling process described below. A tsunami can occur in any tidal state and even at low tide can still **inundate** coastal areas.

G Tsunamis cause damage by two mechanisms: the smashing force of a wall of water travelling at high speed, and the destructive power of a large volume of water draining off the land and carrying a large amount of debris with it, even with waves that do not appear to be large.

H While everyday wind waves have a wavelength (from crest to crest) of about 100m and a height of roughly 2m, a tsunami in the deep ocean has a much larger wavelength of up to 200km. Such a wave travels at well over 800km per hour, but owing to the enormous wavelength the wave oscillation at any given point takes 20 or 30 minutes to complete a cycle and has an amplitude of only about one metre.[2] This makes tsunamis difficult to detect over deep water, where ships are unable to feel their passage. The reason for tsunami, the Japanese name "harbour wave" is that sometimes a village's fishermen would sail out, and encounter no unusual waves while out at sea fishing, and come back to land to find their village **devastated** by a huge wave.

I As the tsunami approaches the coast and the waters become shallow, wave shoaling **compresses** the wave and its speed decreases below 80km per hour. Its wavelength diminishes to less than 20km and its amplitude grows enormously.[3] Since the wave still has the same very long

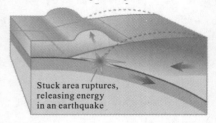

Plate slips, causing subsidence and releasing energy into water

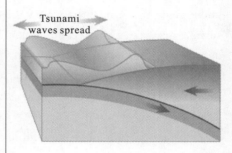

The energy released produces tsunami waves

flexure   *n.* 屈曲,弯曲部分,打褶
amplitude   *n.* 振幅;广阔;角度距离
inundate   *vt.* 淹没;泛滥
compress   *vt.* 压缩;压紧

period, the tsunami may take minutes to reach full height. Except for the very largest tsunamis, the approaching wave does not break, but rather appears like a fast-moving tidal bore. Open bays and coastlines **adjacent** to very deep water may shape the tsunami further into a step-like wave with a steep-breaking front.

J When the tsunami's wave peak reaches the shore, the resulting temporary rise in sea level is termed run up. Run up is measured in metres above a reference sea level. A large tsunami may **feature** multiple waves arriving over a period of hours, with significant time between the wave crests. The first wave to reach the shore may not have the highest run up. (4) About 80% of tsunamis occur in the Pacific Ocean, but they are possible wherever there are large bodies of water, including lakes. They are caused by earthquakes, landslides, volcanic explosions and glacier calvings.

K The understanding of a tsunami's nature remained slim until the 20th century and much remains unknown. Major areas of current research include trying to determine why some large earthquakes do not generate tsunamis while other smaller ones do; trying to accurately forecast the passage of tsunamis across the oceans; and also to forecast how tsunami waves interact with specific shorelines.

adjacent   *adj.* 相邻；邻近的，毗邻的
feature   *n.* 特征，特点
           vt. 使有特色

Notes

(1) Earthquakes, volcanic eruptions, landslides, glacier calvings, and other disturbances above or below water all have the potential to generate a tsunami. 海啸由海底地震、火山爆发、山体滑坡、冰川崩裂及水上或者水下的扰动产生。

(2) Such a wave travels at well over 800km per hour, but owing to the enormous wavelength the wave oscillation at any given point takes 20 or 30 minutes to complete a cycle and has an amplitude of only about one metre. 海啸的波速高达每小时800多千米，因为波长巨大可以在20~30分钟里完成波谷至波峰的震荡，但是在茫茫的大洋里波高却不足1m。

(3) Its wavelength diminishes to less than 20km and its amplitude grows enormously. 当波长减短，少于20km时，波高急剧增高。

(4) The first wave to reach the shore may not have the highest run up. 有时最先到达海岸的浪可能不是最高的浪。

Online Resources

The 2004 Indian Ocean Tsunami

The 2004 Indian Ocean earthquake occurred at 00:58:53 UTC on 26 December with the epicentre off the west coast of Sumatra, Indonesia. The shock had a moment magnitude of 9.1—9.3 and a maximum Mercalli intensity of IX (Violent). The undersea megathrust earthquake was caused when the Indian Plate was subducted by the Burma Plate and triggered a series of devastating tsunamis along the coasts of most landmasses bordering the Indian Ocean, killing 230 000—280 000 people in 14 countries, and inundating coastal communities with waves up to 30m(100ft) high. It was one of the deadliest natural disasters in recorded history. Indonesia was the hardest-hit country, followed by Sri Lanka, India, and Thailand. For more information, please visit https://en.wikipedia.org/wiki/2004_Indian_Ocean_earthquake_and_tsunami.

Exercises

Detailed Understanding

Ⅰ. **Answer the following questions according to the passage you have read.**

1. What may generate a tsunami?
2. Why don't Tsunami waves resemble normal undersea currents or sea waves?
3. Why don't scientists favor that a tsunami is referred to as "tidal waves"?
4. How many people were killed or missing in the 2004 Indian Ocean Tsunami?
5. Why is the term seismic sea wave also used to refer to tsunami? Is seismic sea wave not a completely accurate term?
6. What is the principal generation mechanism of a tsunami?
7. How is a tsunami generated with thrust faults?
8. Explain the two mechanisms by which tsunami cause damage.

Vocabulary

Ⅱ. **Fill in the blanks with the words given below. Change the form where necessary.** (15个单词，10个空)

displacement	wavelength	debris	landslide	basin
seismic	deform	tectonic	equilibrium	amplitude
smashing	compress	devastate	adjacent	shallow

1. The storm caused _____ and flooding in this area.
2. Sedimentary environment and _____ movement have great influence on abundance and thermal evolution extent of organic matters.

Unit 8   Geological Hazard

3. I stood at the foot of the collapsed tower and watched the rescue workers sifting through the _____ .
4. A few days before, a fire had _____ large parts of Windsor Castle.
5. Poor posture, sitting or walking slouched over _____ the body's organs.
6. He sat in an _____ room and waited.
7. Earthquakes produce two types of waves _____ .
8. Sunlight consists of different _____ of radiation.
9. As we fall asleep, the _____ of brain waves slowly becomes greater.
10. Put the milk in a _____ dish.

Sentence Structure

Ⅲ. **Combine each pair of the following sentences, using a "rather than" elanse.**

> **Model**
> A tsunami does not appear as a breaking wave.
> A tsunami may instead initially resemble a rapidly rising tide.
> ——▶Rather than appearing as a breaking wave, a tsunami may instead initially resemble a rapidly rising tide.

1. It exists in dreams.
   It does not exist in actuality.

2. They aim at quality.
   They do not aim at quantity.

3. Boxing is a test of skill and technique.
   Boxing is not brute strength.

4. Let's get it over with as soon as possible.
   Let's not drag it out.

5. You are buying direct.
   You are not buying through an agent.

Translation
Ⅳ. Translate the following sentences into Chinese.
1. Earthquakes, volcanic eruptions, landslides, glacier calvings and other disturbances above or below water all have the potential to generate a tsunami.

2. Tsunami waves do not resemble normal undersea currents or sea waves, because their wavelength is far longer.

3. The 2004 Indian Ocean tsunami was among the deadliest natural disasters in human history with at least 230 000 people killed or missing in 14 countries bordering the Indian Ocean.

4. Prior to the rise of the use of the term tsunami in English-speaking countries, scientists generally encouraged the use of the term seismic sea wave rather than tidal wave.

5. Tsunamis cause damage by two mechanisms: the smashing force of a wall of water travelling at high speed, and the destructive power of a large volume of water draining off the land and carrying a large amount of debris with it, even with waves that do not appear to be large.

Ⅴ. **Translate the following passage into English.**

2004 年印度洋大地震,科学界称为苏门答腊-安达曼地震,发生于 2004 年 12 月 26 日 0 时 58 分 55 秒。印度洋大地震引发的海啸致使 20 多万人丧生,成为历史上最惨重的灾害之一。尽管最终确定震级达到 9.15,仍有许多说法称此次地震的震级为 9.0~9.3,是 1900 年以来规模第二大的地震。2005 年 5 月科学家们声称大部分大地震持续时间仅仅几秒钟,而这次地震几乎持续了 10 分钟。

Reading Skills

Understanding the Sentences with Negative Words

Experienced readers know the fact that it is essential for them to truly understand the sentences with negative words in the reading material. The ability to understand the sentences with negative words is a crucial reading skill.

1. *Unlike* normal ocean waves which are generated by wind, or tides which are generated by the gravitational pull of Moon and Sun, a tsunami is generated by the displacement of water.

2. However, like tsunami, seismic sea wave *is not* a completely accurate term, as forces other than earthquakes—including underwater landslides, volcanic eruptions, underwater explosions, land or ice slumping into the ocean and the weather when the atmospheric pressure changes very rapidly—can generate such waves by displacing water.

3. Tides *do not* play any part in the generation of tsunamis.

4. Except for the very largest tsunamis, the approaching wave *does not* break, but rather appears like a fast-moving tidal bore.

5. The first wave to reach the shore *may not* have the highest run up.

6. The understanding of a tsunami's nature remained slim until the 20th century and much remains *unknown*.

7. Earthquakes can range in size from those that are so weak that they *cannot* be felt to those violent enough to toss people around and destroy whole cities.

8. The sides of a fault move past each other smoothly and aseismically only if there are *no* irregularities or asperities along the fault surface that increase the frictional resistance.

Passage B

Scan and read along the passage

Earthquake[①]

Scan and read along the vocabulary

A An earthquake, also known as a quake, is the shaking of the surface of Earth, resulting from the sudden release of energy in Earth's lithosphere that creates seismic waves. Earthquakes can range in size from those that are so weak that they cannot be felt to those violent enough to toss people around and destroy whole cities. The **seismicity** or seismic activity of an area refers to the frequency, type and size of earthquakes experienced over a period of time.

B Earthquakes are caused mostly by **rupture** of geological faults, but also by other events such as volcanic activity, landslides, mine blasts, and nuclear tests. An earthquake's point of initial rupture is called its focus or **hypocentre**. The **epicentre** is the point at ground level directly above the hypocentre.

C Tectonic earthquakes occur anywhere in Earth where there is sufficient stored **elastic** strain energy to drive **fracture propagation**

seismicity　*n.* 地震活动；受震强度
rupture　*n.* 断裂，破裂
　　　　　vt. 使破裂
hypocentre　*n.* 震源
epicentre　*n.* 震中，中心，集中点
elastic　*adj.* 有弹力的；可伸缩的
fracture　*vt.* 折断，破碎
　　　　　n. 骨折；破裂，断裂
propagation　*n.* 传导

---

①This passage is adapted from: Wikipedia. Earthquake[EB/OL]. (2017-12-29)[2017-12-31]. https://en.wikipedia.org/wiki/Earthquake.

along a fault plane. The sides of a fault move past each other smoothly and aseismically only if there are no irregularities or asperities along the fault surface that increase the frictional resistance. Most fault surfaces do have such asperities and this leads to a form of stick-slip behavior. Once the fault has locked, continued relative motion between the plates leads to increasing stress and therefore, stored strain energy in the volume around the fault surface. This continues until the stress has risen sufficiently to break through the asperity, suddenly allowing sliding over the locked portion of the fault, releasing the stored energy. This energy is released as a combination of radiated elastic strain seismic waves, frictional heating of the fault surface, and cracking of the rock, thus causing an earthquake. This process of gradual build-up of strain and stress punctuated by occasional sudden earthquake fai-lure is referred to as the elastic-rebound theory. It is estimated that only ten percent or less of an earthquake's total energy is radiated as seismic energy. Most of the earthquake's energy is used to power the earthquake fracture growth or is converted into heat generated by friction. Therefore, earthquakes lower Earth's available elastic potential energy and raise its temperature, though these changes are **negligible** compared to the **conductive** and convective flow of heat out from Earth's deep interior.

negligible *adj.* 微不足道的；可以忽略的
conductive *adj.* 传导的
divergent *adj.* 发散的；扩散的
oblique *adj.* 倾斜的；间接的

D There are three main types of fault, all of which may cause an interplate earthquake: normal, reverse (thrust) and strike-slip. Normal and reverse faulting are examples of dip-slip, where the displacement along the fault is in the direction of dip and movement on them involves a vertical component. Normal faults occur mainly in areas where the crust is being extended such as a **divergent** boundary. Reverse faults occur in areas where the crust is being shortened such as at a convergent boundary. Strike-slip faults are steep structures where the two sides of the fault slip horizontally past each other; transform boundaries are a particular type of strike-slip fault. Many earthquakes are caused by movement on faults that have components of both dip-slip and strike-slip; this is known as **oblique** slip.

E Reverse faults, particularly those along convergent plate boundaries are associated with the most powerful earthquakes, megathrust earthquakes, including almost all of those of

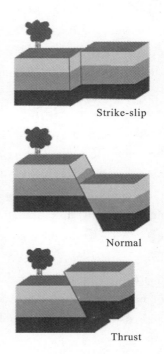

Fault types

magnitude 8 or more. Strike-slip faults, particularly continental transforms, can produce major earthquakes up to about magnitude 8. Earthquakes associated with normal faults are generally less than magnitude 7. For every unit increase in magnitude, there is a roughly **thirtyfold** increase in the energy released. For instance, an earthquake of magnitude 6.0 releases approximately 30 times more energy than a 5.0 magnitude earthquake and a 7.0 magnitude earthquake releases 900 times more energy than a 5.0 magnitude of earthquake. An 8.6 magnitude earthquake releases the same amount of energy as 10 000 atomic bombs like those used in World War II. Strike-slip faults tend to be oriented near vertically, resulting in an approximate width of 10km within the brittle crust, thus earthquakes with magnitudes much larger than 8 are not possible. Maximum magnitudes along many normal faults are even more limited because many of them are located along spreading centres, as in Iceland, where the thickness of the **brittle** layer is only about 6km. In addition, there exists a hierarchy of stress level in the three fault types. Thrust faults are generated by the highest, strike slip by intermediate, and normal faults by the lowest stress levels. This can easily be understood by considering the direction of the greatest principal stress, the direction of the force that "pushes" the rock mass during the faulting. In the case of normal faults, the rock mass is pushed down in a vertical direction, thus the pushing force (greatest principal stress) equals the weight of the rock mass itself. In the case of thrusting, the rock mass "escapes" in the direction of the least principal stress, namely upward, lifting the rock mass up, thus the overburden equals the least principal stress. Strike-slip faulting is intermediate between the other two types described above. This difference in stress **regime** in the three faulting environments can contribute to differences in stress drop during faulting, which contributes to differences in the radiated energy, regardless of fault dimensions.

F An earthquake may cause injury and loss of life, road and bridge damage, general property damage, and collapse or destabilization of buildings. The **aftermath** may bring disease, lack of basic necessities, mental consequences such as panic attacks, depression to survivors and higher insurance **premiums**.

thirtyfold *adj.* 三十倍的
adv. 三十倍地
brittle *adj.* 易碎的
regime *n.* 方法
aftermath *n.* 后果；余波
premium *n.* 保险费

Questions 1—4

Passage B has six paragraphs, A—F

Choose the correct heading for paragraphs B and D—F from the list of headings below.

Write the correct number i — vi in boxes 1—4.

| 1 | | 2 | | 3 | | 4 | |

List of Headings
i. Process of tectonic earthquakes
ii. Dip-slip vs. Strike-slip
iii. Definition of earthquake
iv. Damage and aftermath caused by earthquakes
v. What may cause earthquakes
vi. Fault types: normal, thrust, strike-slip

Example	Paragraph **A**	Answer	iii
1	Paragraph **B**	Answer	
Example	Paragraph **C**	Answer	i
2	Paragraph **D**	Answer	
3	Paragraph **E**	Answer	
4	Paragraph **F**	Answer	

Questions 5—10

Do the following statements reflect the claims of the writer in the passage?

In boxes 5—10, write

  Y(YES)   if the statement reflects the claims of the writer

  N(NO)   if the statement contradicts the claims of the writer

  NG(NOT GIVEN) if it is impossible to say what the writer thinks about this

| 5 | | 6 | | 7 | | 8 | | 9 | | 10 | |

5. Earthquakes are caused only by rupture of geological faults.

6. Most of the earthquake's energy is used to power the earthquake fracture growth or is converted into heat generated by friction.

7. Normal and strike-dip faulting are examples of dip-slip, where the displacement along the fault is in the direction of dip and movement on them involves a vertical component.

8. Many earthquakes are caused by movement on faults that have components of both dip-slip and strike-slip.

9. Strike-slip faults, particularly continental transforms, can produce major earthquakes generally less than magnitude 7.

10. The panic attacks and depression to survivors are the mental consequences only caused by tsunami.

Questions 11－13

Complete the summary of paragraphs E with the list of words A－H below.
Write the correct letter A－H in boxes 11－13.

| 11 | | 12 | | 13 | |

Reverse faults are associated with the most powerful earthquakes including almost all of those of magnitude 8 or more. Strike-slip faults can produce major earthquakes up to about magnitude 8. Earthquakes associated with __11__ faults are generally less than magnitude 7. For every unit increase in magnitude, there is a roughly thirtyfold __12__ in the energy released. There exists a __13__ of stress level in the three fault types. Thrust faults are generated by the highest, strike slip by intermediate, and normal faults by the lowest stress levels.

| A. thrust | B. strike-slip | C. normal | D. hierarchy |
| E. increase | F. lowest | G. stress | H. intermediate |

Glossary

Unit 1

adversely /æd'vɜːsli/
　adv. 不利地;逆地;反对地
altitude /ˈæltɪtjuːd/
　n. (地表面、海平面之上的)高;高度;(尤指)海拔;深(度);垂直距离;平地纬度
appreciably /əˈpriːʃəbli/
　adv. 明显地;相当地;可察觉地
arbitrary /ˈɑːbɪtrəri/
　adj. 任意的;武断的
astronomy /əˈstrɒnəmi/
　n. 天文学
astrophysics /ˌæstrəʊˈfɪzɪks/
　n. 天体物理学
atmosphere /ˈætməsfɪə(r)/
　n. 气氛;大气;空气;大气圈;大气层
atmospheric chemistry
　n. 大气化学
atop /əˈtɒp/
　prep. 在…的顶上
　adv. 在顶上
biosphere /ˈbaɪəsfɪə(r)/
　n. 生物圈
canyon /ˈkænjən/
　n. 峡谷
catastrophism /kəˈtæstrəfɪzəm/
　n. 灾变说;劫数难逃论
chronological /ˌkrɒnəˈlɒdʒɪkəl/
　adj. 按年代顺序排列的;依时间前后排列记载的

climatology /ˌklaɪməˈtɒlədʒi/
　n. 〈气候〉气候学;风土学
composition /ˌkɒmpəˈzɪʃn/
　n. 作文,作曲;构成;合成物;组成
core /kɔː/
　n. 地核
crust /krʌst/
　n. 外壳;面包皮;坚硬外皮
　vi. 结硬皮;结成外壳
　vt. 盖以硬皮;在…上结硬皮
doctrine /ˈdɒktrɪn/
　n. 主义;学说;教义;信条
erosion /ɪˈrəʊʒən/
　n. 侵蚀,腐蚀
evaporate /ɪˈvæpəreɪt/
　vt. 使……蒸发;使……脱水;使……消失
　vi. 蒸发,挥发;消失,失踪
excavation /ˌekskəˈveɪʃn/
　n. 挖掘,发掘
extraction /ɪkˈstrækʃn/
　n. 提取,抽出;精炼
forecast /ˈfɔːkɑːst/
　vt. 预报,预测;预示
　vi. 进行预报,作预测
　n. 预测,预报;预想
fossil /ˈfɒsl/
　n. 化石;僵化的事物;顽固不化的人
　adj. 化石的;陈腐的,守旧的
fraction /ˈfrækʃn/
　n. 分数;部分;小部分;稍微
freshwater /ˈfreʃwɔːtə(r)/
　n. 淡水;内河;湖水

galaxy /ˈgæləksi/
    n. 银河；星系；银河系；一群显赫的人
gem /dʒem/
    n. 宝石
geochemistry /ˌdʒiəʊˈkemistri/
    n. 地球化学
geosphere /ˈdʒiːəʊsfiə(r)/
    n. 岩石圈；陆界
glacial /ˈgleisiəl；—ʃəl/
    adj. 冰的；冰冷的；冰河时代的
granodiorite /ˌgrænəʊˈdaiərait/
    n. 花岗闪长岩
groundwater /ˈgraʊndˌwɔːtə(r)/
    n. 地下水
hazard /ˈhæzəd/
    n. 危险；灾害
historical geology
    n. 地史学
holistic /həʊˈlistik/
    adj. 整体的；全盘的
homogeneous /ˌhɒmə(u)ˈdʒiːniəs/
    adj. 均质的；同性质的；同类的；同种的；齐次的；相同特征的
hydrosphere /ˈhaidrəsfiə(r)/
    n. 水界；水圈
igneous /ˈigniəs/
    adj. 火的；似火的；火熔的；(尤指岩石)火成的；igneous rock 火成岩
inhabit /inˈhæbit/
    vt. 栖息；居住于；占据；
    vi. 居住；栖息
innumerable /iˈnjuːmərəbl/
    adj. 无数的，不计其数的，数不清的
integrate /ˈintigreit/
    vt. 使……完整；使……成整体；求……的积分
    vi. 求积分；取消隔离；成为一体
    adj. 整合的；完全的
    n. 一体化；集成体
interaction /ˌintərˈækʃn/
    n. 相互作用；交互作用；互动
interconnected /ˌintəkəˈnektid/
    adj. 连通的；有联系的
interdependent /ˌintədiˈpendənt/
    adj. 相互依赖的；互助的
interdisciplinary /ˌintəˈdisiplinəri/
    adj. 各学科间的
iron-nickel alloy
    n. 铁镍合金
landscape /ˈlændskeip/
    n. 风景，山水画；地形
landslide /ˈlændslaid/
    n. 山崩；滑坡
lens /lenz/
    n. 镜头；透镜；晶状体
literally /ˈlitərəli/
    adv. 照字面地；逐字地；不夸张地
lunar /ˈljuːnə(r)/
    adj. 月亮的，月球的；阴历的
magnetic /mægˈnetik/
    adj. 地磁的；有磁性的；有吸引力的
magnitude /ˈmægnitjuːd/
    n. 大小；量级；〈地震〉震级；重要；光度
mantle /ˈmæntl/
    n. 地幔；斗篷；覆盖物
marine geology
    n. 海洋地质学（亦作 geological oceanography）
megacity /ˈmegəsiti/
    n. （人口超过千万的）大城市；巨型城市；特大都市
metallic /miˈtælik/
    adj. 金属的，含金属的
meteorite /ˈmiːtiərait/
    n. 陨星；流星
meteorology /ˌmiːtiəˈrɒlədʒi/
    n. 气象状态，气象学
mountain range
    n. 山脉；山岳地带，山地

Glossary

multitude /ˈmʌltitjuːd/
  n. 群众；多数
numerous /ˈnjuːmərəs/
  adj. 许多的，很多的
oceanography /ˌəuʃəˈnɔgrəfi/
  n. 海洋学
organism /ˈɔːgənizəm/
  n. 有机体；生物体；微生物
palaeontology /ˌpæliɔnˈtɔlədʒi/
  n. 古生物学
physical geology
  n. 普通地质学
physical oceanography
  n. 海洋物理学
pillar /ˈpilə(r)/
  n. 柱子，栋梁，墩；支柱
planetary science
  n. 行星科学
precede /priˈsiːd/
  vt. 领先，在……之前；优于，高于；
  vi. 领先，在前面
precipitate /priˈsipiteit/
  vt. 使沉淀；促成；
  vi. 沉淀；猛地落下；冷凝成为雨或雪等
prominent /ˈprɔminənt/
  adj. 突出的，显著的；杰出的；卓越的
radius /ˈreidiəs/
  n. 半径，半径范围；界限
salt water
  n. 盐水；海水
scupt /skʌlpt/
  v. 造型，雕刻，使成形
seismic /ˈsaizmik/
  adj. 地震的；因地震而引起的
silicon /ˈsilikən/
  n. 硅；硅元素
staggering /ˈstægəriŋ/
  adj. 惊人的，令人震惊的

sulfur /ˈsʌlfə(r)/
  vt. 用硫磺处理
  n. 硫磺；硫磺色
survival /səˈvaivəl/
  n. 幸存，残存；幸存者，残存物
tenet /ˈtenit/
  n. 原则；信条；教义
the Himalayas
  n. 喜马拉雅山脉
the Rockies
  n. 落基山脉，纵贯加拿大、美国的山脉
thwart /θwɔːt/
  vt. 挫败；反对；阻碍
ultraviolet radiation
  n. 紫外线；紫外线照射
uniformitarianism /ˌjuːnifɔːmiˈteəriənizəm/
  n. 均变论；推今及古原理
unravel /ʌnˈrævl/
  vt. 解开，阐明；解决；拆散
  vi. 解决；散开
urbanisation /ˌəːbənaiˈzeiʃn/
  n. 都市化；城市化
viable /ˈvaiəbl/
  adj. 可行的；能活的；有生活力的
vulnerable /ˈvʌlnərəbl/
  adj. 易受攻击的，易受伤害的；有弱点的；脆弱的
waste /weist/
  n. 浪费；废物；荒地；损耗；地面风化物

Unit 2

abundance /əˈbʌndəns/
  n. 丰度；丰富，充裕；大量，极多；盈余
accumulate /əˈkjuːmjəleit/
  vt. 堆积，积累
  vi. (数量)逐渐增加，(质量)渐渐提高
alarm /əˈlɑːm/
  n. 警报；惊恐

v. 警告；使惊恐
analytical /ˌænəˈlitikl/
 adj. 分析的，分析法的
argon /ˈɑːɡɔn/
 n. 氩
aspire /əˈspaiə(r)/
 v. 渴望；立志；追求
atomic /əˈtɔmik/
 adj. 原子的；原子能的，原子武器的；极微的
awful /ˈɔːfl/
 adj. 可怕的；糟糕的；非常的
benchmark /ˈbentʃmɑːk/
 n. 基准，参照；标准检查程序
biase /baiəs/
 n. 偏差
bivalve /ˈbaivælv/
 n. 双壳类；双阀
catalogue /ˈkætəlɔɡ/
 n. 目录，一览表
catastrophic /ˌkætəˈstrɔfik/
 n. 灾难的；惨重的
chronological /ˌkrɔnəˈlɔdʒikl/
 adj. 按时间的前后顺序排列的；编年的
colleague /ˈkɔliːɡ/
 n. 同事；同行
convection /kənˈvekʃn/
 n. 传送；对流；运流
correlate /ˈkɔrəleit/
 v. 联系；使互相关联
 n. 相关物；相关联的人
 adj. 相关的；相应特点的
curve /kəːv/
 n. 弧线，曲线；曲线状物
 v. 使弯曲；使成曲线；使成弧形
database /ˈdeitəbeis/
 n. 数据库；资料库
decay /diˈkei/
 n. 腐败；衰退的状态
 vt. 衰败，衰退，衰落

 vi. 腐烂，腐朽
deposit /diˈpɔzit/
 vi. 沉淀
 vt. 储蓄；寄存；放置，安置；付保证金
 n. 保证金；存储；储蓄，存款，押金；沉淀物；堆积物；寄存，寄存品；化石；僵化的事物；顽固不化的人
 adj. 化石的；陈腐的；守旧的
dislodge /disˈlɔdʒ/
 vi. 移走，离开原位
 vt. 把……逐出，驱逐；把……移动
doomsday /ˈduːmzdei/
 n. 世界末日；遭灾之日
duplicate /ˈdjuːplikeit/
 v. 重复；复制
 adj. 复制的；副本的
 n. 副本；完全一样的东西
ecosystem /ˈiːkəusistəm/
 n. 〈生〉生态系统
elapse /iˈlæps/
 vi. 消逝；时间过去
 n. （时间的）消逝
episode /ˈepisəud/
 n. 插曲；一集；片段
equivalent /iˈkwivələnt/
 adj. 相等的，相当的
evolve /iˈvɔlv/
 v. 发展；进化；设计
extinction /ikˈstiŋkʃn/
 n. 熄灭；消灭；灭绝
faunal /ˈfɔːnl/
 adj. 动物区系的
fungi /ˈfʌŋɡiː/
 n. (fungus的复数)真菌；霉，霉菌
gigantic /dʒaiˈɡæntik/
 adj. 巨大的，庞大的
horizontality /hɔrizɔnˈtæləti/
 n. 水平状态（或位置、性质）
imperfection /ˌimpəˈfekʃn/

  n. 不完美；缺点，瑕疵
impress /imˈpres/
  v. 印；给……以深刻印象
  n. 印象；印记
inadequate /inˈædikwət/
  adj. 不充足的；不适当的；不足胜任的
incorporate /inˈkɔːpəreit/
  vt. 包含；组成公司；使混合
  vi. 合并；包含；吸收；混合
interval /ˈintəvl/
  n. 间隔；幕间休息；区间
isotope /ˈaisətəup/
  n. 〈化〉同位素
jellyfish /ˈdʒelifiʃ/
  n. 水母；海蜇；软弱无用的人；意志薄弱的人
literature /ˈlitrətʃə(r)/
  n. 文学；文献；文学作品
lust /lʌst/
  n. 欲望；渴求
  v. 贪求；渴望
magnet /ˈmægnət/
  n. 磁铁，磁石；有吸引力的人或物；〈物〉磁体
mammal /ˈmæml/
  n. 哺乳动物
marine /məˈriːn/
  adj. 海的；海产的；海军的
  n. 水兵；海军士兵；海事，海运业
mathematical /mæθəˈmætikl/
  adj. 数学的；精确的；绝对的；可能性极小的
mimic /ˈmimik/
  v. 摹拟；模仿
  n. 巧于模仿的人；复写品/仿制品
  adj. 模仿的，摹拟的
multi-storey /mʌlˈtistˈɔːri/
  adj. 多层的
mystery /ˈmistri/
  n. 秘密，谜；神秘
neutron /ˈnjuːtrɔn/
  n. 〈物〉中子

niche /nitʃ/
  n. 壁龛；合适的位置（工作等）；商机
nucleus /ˈnjuːkliəs/
  n. 中心，核心；（原子）核；起点，开始；〈微〉细胞核
palaeobiology /ˌpæliəubaiˈɔlədʒi/
  n. 古生物学
palaeomagnetism /ˌpæliəuˈmægnitizəm/
  n. 古地磁；古磁学
patchy /ˈpætʃi/
  adj. 不调和的；拼凑成的；有补丁的
plateau /ˈplætəu/
  n. 高原；平稳时期；停滞期
polarity /pəˈlærəti/
  n. 〈物〉极性；〈生〉反向性；对立；〈数〉配极
potassium /pəˈtæsiəm/
  n. 〈化〉钾
proton /ˈprəutɔn/
  n. 〈物〉质子
radiation /reidiˈeiʃn/
  n. 辐射；放射物；辐射状；分散
radioactive /ˌreidiəuˈæktiv/
  adj. 放射性的
radiometric /ˌreidiəuˈmetrik/
  adj. 辐射度的，放射性测量的
repository /riˈpɔzətri/
  n. 仓库；贮藏室；博物馆
resonance /ˈrezənəns/
  n. 共振；共鸣；反响
reversal /riˈvəːsl/
  n. 倒转，颠倒；反复；逆转，反转
rival /ˈraivl/
  n. 对手；竞争者
  v. 竞争；比得上
scenario /səˈnɑːriəu/
  n. 方案；剧情概要；分镜头剧本
sceptical /ˈskeptikl/
  adj. 怀疑的；怀疑论者的
sedimentary /ˌsediˈmentri/
  adj. 沉积的，沉淀性的

strata /ˈstrɑːtə/
  n. 层；〈地质〉地层；岩层（stratum 的名词复数）；社会阶层
substitute /ˈsʌbstitjuːt/
  v. 代替，替换
  n. 替代物；代替者；替补
succession /səkˈseʃn/
  n. 〈生〉自然演替；一系列，接连；继承人，继承权，继承顺序
superposition /ˌsjuːpəpəˈziʃn/
  n. 叠加；叠置；叠覆
taxonomist /tækˈsɔnəmist/
  n. 分类学者
thoroughly /ˈθʌrəli/
  adv. 彻底地；完全地
trilobite /ˈtrailəubait/
  n. 三叶虫
undeformed /ˌʌndiˈfɔːmd/
  adj. 无形变的
undeniably /ˌʌndiˈnaiəbli/
  adv. 不可否认地
undertaking /ˌʌndəˈteikiŋ/
  n. 事业；保证；企业
urchin /ˈəːtʃin/
  n. 顽童；淘气鬼；〈主英〉〈动〉猬；〈动〉海胆
volcanic /vɔlˈkænik/
  adj. 火山的；猛烈的；暴烈的
  n. 火山岩
wizardry /ˈwizədri/
  n. 魔法；魔术；杰出才能
wrangle /ˈræŋgl/
  v. 争吵；争论

Unit 3

adjoining /əˈdʒɔiniŋ/
  adj. 邻接的；毗连的；
  v. 邻接（adjoin 的 ing 形式）
associate with
  v. 联合；与……联系在一起；和……来往
asthenosphere /əsˈθenəsfiə/
  n. 软流圈；岩流圈
astronomical /ˌæstrəˈnɔmikl/
  adj. 天文的，天文学的；极大的
boundary /ˈbaundəri/
  n. 边界；范围；分界线
buoyant /ˈbɔiənt/
  adj. 轻快的；有浮力的；上涨的
compelling /kəmˈpeliŋ/
  adj. 引人入胜的，扣人心弦的，非常强烈的，不可抗拒的（compel v. 强迫；使不得不）
continental volcanic arc
  n. 火山弧（板块俯冲过程形成的一串火山，从上看其外形像个弧）
continuity /ˌkɔntiˈnjuːiti/
  n. 连续性；一连串
deform /diˈfɔːm/
  vt. 使变形；使成畸形；使残废
differentiate /ˌdifəˈrenʃieit/
  vi. & vt. 区分；区别
embarked on
  v. 从事，着手；登上船
expedition /ˌekspəˈdiʃn/
  n. 远征；探险队；迅速
fault /fɔːlt; fɔlt/
  n. 故障；错误；缺点；毛病；〈地质〉断层（normal fault 正断层）
  vi. 弄错；〈地质〉产生断层
fault-block mountains
  n. 断块山地
fault-fold mountains
  n. 断褶山地
fold /fəuld/
  n. 褶皱
geophysicist /ˌdʒiːəuˈfizisist/
  n. 地球物理学者
grind /graind/
  vt. 磨碎；磨快

vi. 磨碎；折磨
hemisphere /ˈhemisfiə/
　n. 半球
hypothesis /haiˈpɔθisis/
　n. 假设（hypothesise〈英〉/hypothesize〈美〉
　v. 假设；假定；设定）
landmass /ˈlændmæs/
　n. 大陆
lithosphere /ˈliθəsfiə(r)/
　n. 陆界；岩石圈
mechanism /ˈmekənizəm/
　n. 机制；原理，途径；机械装置；技巧
meteorologist /ˌmiːtiəˈrɔlədʒist/
　n. 气象学者
oceanic ridge
　n. 洋脊；海岭（mid-ocean ridge 洋中脊）
palaeoclimatic /ˌpæliəˌklaiˈmætik/
　adj. 古气候的（palaeo = acient, climatic = climate）
Palaeozoic /ˌpæliːəuˈzəuik/
　adj. 古生代的
Pangaea /pænˈdʒiːə/
　n. all lands 泛大陆；盘古大陆
release /riˈliːs/
　vt. 释放；发射；让与；允许发表；
　n. 释放；发布；让与
reptile /ˈreptail/
　adj. 爬虫类的；
　n. 爬行动物
sediment /ˈsedimənt/
　n. 沉积；沉淀物（如沙、泥等）
speculate /ˈspekjuleit/
　vi. 推测；投机；思索
　vt. 推断；投机；思索
subduction /səbˈdʌkʃn/
　n. 俯冲；除去；减法
supercontinent /ˈsuːpəkɔntinənt/
　n. 超大陆

swamp /swɔmp/
　n. 湿地；沼泽
tectonic /tekˈtɔnik/
　adj. 构造的；建筑的；地壳构造上的（tectonic activity 构造活动）
thermal /ˈθəːməl/
　adj. 热的；热量的；保热的
tilt /tilt/
　vi. 倾斜；翘起；以言词或文字抨击；
　vt. 使倾斜；使翘起；
　n. 倾斜
trench /trentʃ/
　n. 沟，沟渠；战壕（deep-ocean trench 深海沟）
unprecedented /ʌnˈpresidentid/
　adj. 空前的；无前例的
upwell /ʌpˈwel/
　vi. 上涌；往上涌出
vibration /vaiˈbreiʃn/
　n. 振动

Unit 4

abiogenic /eibaiəudˈʒenik/
　adj. 自然发生的
acicular /əˈsikjulə/
　adj. 针状的，针尖状的
adamantine /ˌædəˈmæntain/
　adj. 非常坚硬的；
　n. 金刚合金
aggregate /ˈægrigit/
　n. 合计；聚集体
allochromatic /æləkrəˈmætik/
　adj. 无色但含有色杂质的矿物的
anhedral /ˈænhiːdrəl/
　n. 上反角
anthropogenic /ˌænθrəpəˈdʒnik/
　adj. 人为的；〈人类〉人类起源的，人类活动产生的

apatite /ˈæpətait/
　　n. 磷灰石
apophyllite /əˈpɒfilait/
　　n. 鱼眼石
asbestiform /æzˈbestifɔːm/
　　adj. 石棉状的
azurite /ˈæʒurait/
　　n. 石青；蓝铜矿
basalt /bəˈsɔːlt/
　　n. 玄武岩
biogenic /ˌbaiəuˈdʒenik/
　　adj. 源于生物的，生物所造成的
biotite /ˈbaiətait/
　　n. 黑云母
bornite /ˈbɔːnait/
　　n. 斑铜矿
calcite /ˈkælsait/
　　n. 方解石
calculi /ˈkælkjəlai/
　　n. 结石；积石
chrysotile /ˈkrisə(u)tail/
　　n. 温石绒，温石棉
clastic /ˌklæstik/
　　adj. 可分解的，碎屑状的
corundum /kəˈrʌndəm/
　　n. 刚玉，金刚砂
crystalline /ˈkristəlain/
　　adj. 水晶的；结晶质的
　　n. 结晶性，结晶度
crystallographic /ˌkristələˈgræfik/
　　adj. 晶体的；晶体学的
dendritic /denˈdritik/
　　adj. 树枝状的
detritus /diˈtraitəs/
　　n. 碎石；(侵蚀形成的)岩屑；风化物
diagnostic /daiəgˈnɒstik/
　　adj. 诊断的，判断的；特征的
diaphaneity /ˌdaiəfəˈniːəti/
　　n. 透明度；透明性

diffraction /diˈfrækʃn/
　　n. 衍射，宽龟裂状的
diorite /ˈdaiərait/
　　n. 闪长岩
dunite /ˈdʌnait/
　　n. 纯橄榄岩
elongate /iˈlɔːŋgeit/
　　vt. 延长，加长
encompass /ənˈkʌmpəs/
　　vt. 包含；包括
equant /i(ː)ˈkwɒnt/
　　adj. 等径的，等分的
equivocation /iˌkwivəˈkeiʃn/
　　n. 模棱两可的话，含糊话
euhedral /juːˈhiːdrəl/
　　adj. 自形的
exclude /ikˈskluːd/
　　vt. 排斥；排除，不包括
extrusive /ikˈstruːsiv/
　　adj. 挤出的，喷出的，突出的
feldspar /ˈfəldpɑː/
　　n. 长石
fibrous /ˈfaibrəs/
　　adj. 含纤维的，纤维性的
fluorite /ˈfluːərait/
　　n. 萤石，氟石
foliated /ˈfəulieitid/
　　v. 有叶的，覆有叶的
formula /ˈfɔːrmjələ/
　　n. 公式；方案
gabbro /ˈgæbrəu/
　　n. 辉长岩
galena /gəˈliːnə/
　　n. 方铅矿
garnet /ˈgɑːrnit/
　　n. 石榴石
genesis /ˈdʒenisis/
　　n. 创始，起源，发生
geometric /ˌdʒiəˈmətrik/

adj. 几何学的
gneiss /nais/
   n. 片麻岩
granite /ˈɡrænit/
   n. 花岗岩，花岗石
granodiorite /ˌɡrænəuˈdaiərait/
   n. 花岗闪长岩
granular /ˈɡrænjələ(r)/
   adj. 颗粒状的
gypsum /ˈdʒipsəm/
   n. 石膏
hematite /ˈhemətait/
   n. 赤铁矿
hornfels /ˈhɔrnfls/
   n. 角页岩
idiochromatic /ˈidiəukrəuˈmætik/
   adj. 自色的；本质色
incandescence /ˌinkənˈdəsəns/
   n. 白炽；白热
intrusion /inˈtruːʒn/
   n. 干扰，干涉
intrusive /inˈtrusiv/
   adj. 侵入的；闯入的，打扰的
ionic /aiˈɑːnik/
   adj. 离子的
jadeite /ˈdʒeidait/
   n. 硬玉；翠
kyanite /ˈkaiəˌnait/
   n. 蓝晶石
labradorite /ˌlæbrəˈdɔrait/
   n. 拉长岩
lava /ˈlɑːvə/
   n. 熔岩；火山岩
lustre /ˈlʌstə(r)/
   n. 光泽；光彩
macroscopic /ˌmækrəˈskɔpik/
   adj. 宏观的；肉眼可见的
mafic /ˈmæfik/
   adj. 铁镁质的

magma /ˈmæɡmə/
   n. 岩浆
malachite /ˈmæləkait/
   n. 孔雀石
marble /ˈmɑːbl/
   n. 大理石
metamorphic /ˌmetəˈmɔːfik/
   adj. 变形的，变质的，改变结构的
metamorphism /metəˈmɔːfizəm/
   n. 变质；变形
mineraloid /ˈminərəˌlɔid/
   n. 准矿物；似矿物
muscovite /ˈmʌskvait/
   n. 白云母
mylonite /ˈmailəˌnait/
   n. 糜棱岩
natrolite /ˈnætrəlait/
   n. 钠沸石
nephrite /ˈnefrˌait/
   n. 软玉
opaque /əuˈpeik/
   adj. 不透明的；无光泽的，晦暗的
orthoclase /ˈɔːθəkleis/
   n. 正长石
oxalate /ˈɔksəleit/
   n. 草酸盐
peridotite /ˌperiˈdəutait/
   n. 橄榄岩
periodic /ˌpiəriˈɔdik/
   adj. 周期的；定期的；间歇的
permeability /pəˌmiəˈbiliti/
   n. 渗透性；磁导率
petrographic /ˌpetrəuˈɡræfik/
   adj. 岩相学的，岩类学的
petrography /piˈtrɔɡrəfi/
   n. 岩石记述学
petrology /pəˈtrɑːlədʒi/
   n. 岩石学

phyllite /ˈfilait/
  n. 千枚岩,硬绿泥石
phyllosilicate /filəˈsilikeit/
  n. 页硅酸盐
platy /ˈpleiti/
  adj. 〈岩〉裂成平坦薄片的,板状的,扁平状的
plutonic /pluːˈtɔnik/
  adj. 深成的,(岩石)火成的
polymorph /ˈpɔlimɔːf/
  n. 多晶型物,多形态的动(植)物
prismatic /prizˈmætik/
  adj. 棱形,棱镜的
protolith /prəutəuliθ/
  n. 原岩
pumice /ˈpʌmis/
  n. 轻石;浮石
pyrite /ˈpairait/
  n. 黄铁矿
qualitative /ˈkwɑːləteitətiv/
  adj. 定性的,定质的;性质上的
quartz /kwɔːrts/
  n. 石英
reflectivity /ˌriːflekˈtiviti/
  n. 反射率
resinous /ˈrezinəs/
  adj. 树脂质的,含树脂的
ruby /ˈrubi/
  n. 红宝石
sapphire /ˈsæfaiə/
  n. 蓝宝石
shale /ʃeil/
  n. 〈岩〉页岩;泥板岩
silicate /ˈsilikeit/
  n. 硅酸盐
siltstone /ˈsiltˌstəun/
  n. 粉砂岩
slate /sleit/
  n. 石板;板岩,页岩

soapstone /ˈsəupstəun/
  n. 皂石;滑石
subhedral /sʌbˈhiːdrəl/
  adj. 半形的,仅有部分晶面的
syenites /ˈsaiənait/
  n. 正长岩
symmetry /ˈsimitri/
  n. 对称; 匀称
synonymous /siˈnɔniməs/
  adj. 同义词的;同义的
tabular /ˈtæbjələ(r)/
  adj. 表格的;平坦的;平板的
talc /tælk/
  n. 滑石,云母
topaz /ˈtəupæz/
  n. 黄玉,托帕石
translucent /trænsˈluːs(ə)nt/
  adj. 半透明的;透亮的,有光泽的
transparent /trænsˈpærənt/
  adj. 透明的;清澈的
underlying /ˌʌndərˈlaiiŋ/
  adj. 潜在的;基础的;表面下的
urinary /ˈjuərn(ə)ri/
  adj. 尿的;泌尿的
vitreous /ˈvitriəs/
  adj. 玻璃(似)的,玻璃质的
  n. 玻璃状态,透明性

Unit 5

accumulate /əˈkjuːmjuleit/
  v. 聚积
aquaculture /æˈkwəkʌltʃə(r)/
  n. 水产养殖业
arch /ɑːtʃ/
  n. 背斜;穹隆;天生桥
  vt. 使……弯成弓形;用拱连接
  vi. 拱起;成为弓形

Glossary

artificial /ˌɑːtɪˈfɪʃl/
    adj. 人造的
astound /əsˈtaʊnd/
    v. 使惊讶
buff /bʌf/
    vt. 记有软皮摩擦；缓冲；擦亮，抛光某物
canal /kəˈnæl/
    n. 水道
chisel /ˈtʃɪzl/
    vt. 雕，刻；凿；欺骗
    vi. 雕，刻；凿；欺骗
    n. 砾石滩
concur /kənˈkɜː(r)/
    v. 同意
copper /ˈkɒpə(r)/
    n. 铜
crevice /ˈkrevɪs/
    n. 裂缝；裂隙
crumble /ˈkrʌmbl/
    vi. 崩溃；破碎，崩解
    vt. 崩溃；弄碎，粉碎
decomposed /ˌdiːkəmˈpəʊzd/
    adj. 已腐烂的，已分解的
decomposition /ˌdiːkɒmpəˈzɪʃn/
    n. 分解，腐败
delta /ˈdeltə/
    n. 三角洲
desalination /diːˌsælɪˈneɪʃn/
    n. 海水淡化
devise /dɪˈvaɪz/
    n. 设计
divert /daɪˈvɜːt/
    v. 转移
dune /djuːn/
    n. （由风吹积而成的）沙丘
erode /ɪˈrəʊd/
    v. 腐蚀，侵蚀，水土流失
exogenetic /ˈeksəʊdʒɪˈnetɪk/
    adj. 外生的；外因的；外源性的

fertile /ˈfɜːtaɪl/
    adj. 肥沃的
fertiliser /ˈfɜːtəlaɪzə(r)/
    n. 肥料
frigid /ˈfrɪdʒɪd/
    adj. 寒冷的，严寒的
indicate /ˈɪndɪkeɪt/
    v. 表明
interior /ɪnˈtɪəriə(r)/
    n. 内陆
investigation /ɪnˌvestɪˈgeɪʃn/
    n. 调查
irrigation /ˌɪrɪˈgeɪʃn/
    n. 灌溉
karst /kɑːst/
    n. 喀斯特；岩溶
lagoon /ləˈguːn/
    n. 潟湖
lead /led/
    n. 铅
Mediterranean /ˌmedɪtəˈreɪniən/
    adj. 地中海的
mercury /ˈmɜːkjʊri/
    n. 汞
moraine /məˈreɪn/
    n. 冰碛；（熔岩流表面的）火山碎屑
nutrient /ˈnjuːtriənt/
    n. 养料
oxidation /ˌɒksɪˈdeɪʃn/
    n. 〈化〉氧化
parched /pɑːtʃt/
    adj. 焦的；炎热的；炒过的；干透的
    vt. 烘干，使极渴（parch 的过去分词）
pebble /ˈpebl/
    n. 中砾，卵石
    v. （用卵石等）铺
pockmark /ˈpɒkmɑːk/
    n. 麻子；凹坑

reduce /rɪˈdjuːs/
 vt. 缩减;简化;还原
remnant /ˈremnənt/
 n. 剩余
 adj. 剩余的
replenish /rɪˈplenɪʃ/
 v. 补充
scour /ˈskaʊə(r)/
 v. 冲刷
sinkhole /ˈsɪŋkhəʊl/
 n. 落水洞;灰岩坑
slit /slɪt/
 n. 淤泥
stimulate /ˈstɪmjʊleɪt/
 v. 刺激
towering /ˈtaʊərɪŋ/
 adj. 高耸的;卓越的;激烈的
virtually /ˈvɜːtʃʊəli/
 adv. 差不多
waterway /ˈwɔːtəweɪ/
 n. 水道
wedge /wedʒ/
 n. 楔形体
zinc /zɪŋk/
 n. 锌

Unit 6

abruptly /əˈbrʌptli/
 adv. 突然地;唐突地
acidification /əˌsɪdɪfɪˈkeɪʃn/
 n. 〈化〉酸化;成酸性;使……发酸
airborne /ˈeəbɔːn/
 adj. 〈航〉空运的;空气传播的;风媒的
albedo /ælˈbiːdəʊ/
 n. (行星等的)反射率;星体反照率(复数:albedos)
amplify /ˈæmplɪfaɪ/
 v. 放大;详述
 adj. 放大的;扩充的
anticipate /ænˈtɪsɪpeɪt/
 vt. 预期,期望;占先,抢先;提前使用
arctic /ˈɑːktɪk/
 adj. 北极的;极寒的
 n. 北极圈;御寒防水套鞋
collaborator /kəˈlæbəreɪtə(r)/
 n. 〈劳经〉合作者;勾结者;通敌者
curb /kɜːb/
 n. 抑制;路边;勒马绳
 vt. 控制;勒住
devastate /ˈdevəsteɪt/
 vt. 毁灭;毁坏
dimmer /ˈdɪmə(r)/
 n. (车辆)调光器;光暗掣;衰减器
dwarf /dwɔːf/
 adj. 矮小的
 n. 侏儒,矮子
 vi. 变矮小
 vt. 使矮小
elaborate /ɪˈlæbərət/
 adj. 精心制作的;详尽的;煞费苦心的
 vt. 精心制作;详细阐述;从简单成分合成(复杂有机物)
 vi. 详细描述;变复杂
ellipse /ɪˈlɪps/
 n. 〈数〉椭圆形;椭圆
external /ɪkˈstɜːnl/
 adj. 外部的;表面的;〈药〉外用的;外国的;外面的
flip /ˈflɪp/
 vt. 轻击;急挥
 vi. 发疯;捻;蹦蹦跳跳
 n. 浏览;空翻;跳跃
 adj. 无礼的;冒失的
 n. 振动;波动;动摇;〈物〉振荡

implausible /ɪmˈplɔːzɪbl/
 adj. 难以置信的,不像真实的
induce /ɪnˈdjuːs/
 vt. 诱导;引起;引诱;感应
infrared /ˌɪnfrəˈred/
 n. 红外线
 adj. 红外线的
internal /ɪnˈtɜːnl/
 n. 内脏;本质
 adj. 内部的;里面的;体内的;(机构)内部的
levee /ˈlevi/
 n. 堤坝(码头);(旧时君主或显贵的)早晨接见
 vt. 为……筑堤
mean /miːn/
 n. 平均值
 adj. 平均的;卑鄙的;低劣的;吝啬的
 vi. 用意
 vt. 意味;想要;意欲
meridional /məˈrɪdiənl/
 adj. 南欧的;子午线的;南部的
 n. 南欧人
oscillation /ˌɒsɪˈleɪʃn/
plasma /ˈplæzmə/
 n. 〈等离子〉等离子体;血浆;〈矿物〉深绿玉髓
property /ˈprɒpəti/
 n. 性质,性能;财产;所有权
spectrum /ˈspektrəm/
 n. 光谱;频谱;范围;余象(复数: spectrums, spectra)
statistical /stəˈtɪstɪkl/
 adj. 统计的;统计学的
target /ˈtɑːɡɪt/
 n. 目标;靶子
 vt. 把……作为目标;规定……的指标;瞄准某物
thermal /ˈθɜːml/
 adj. 热的;热量的;保热的
 n. 上升的热气流
tributary /ˈtrɪbjətri/
 adj. 纳贡的;附属的;辅助的
 n. 支流;进贡国;附属国
utterly /ˈʌtəli/
 adv. 完全地;绝对地;全然地;彻底地,十足地
variation /ˌveəriˈeɪʃn/
 n. 变化;〈生〉变异,变种

Unit 7

ailing /ˈeɪlɪŋ/
 adj. 生病的;不舒服的
algae /ˈældʒiː/
 n. 〈植〉藻类;〈植〉海藻
anhydrite /ænˈhaɪdraɪt/
 n. 〈矿物〉硬石膏;〈建〉无水石膏
bacterial /bækˈtɪəriəl/
 adj. 〈微〉细菌的
blade /bleɪd/
 n. 桨叶;刀片,剑;(壳、草等的)叶片
brackish /ˈbrækɪʃ/
 adj. 含盐的
brine /braɪn/
 n. 卤水;盐水;海水
bubble /ˈbʌbl/
 n. 泡,水泡;冒泡,起泡,泡影,妄想
 v. (使)冒泡
cable /ˈkeɪbl/
 n. 缆绳,绳索
 v. 发电报至;电传;固定系牢
calcareous /kælˈkeəriəs/
 adj. 钙质的,石灰质的
canalization /ˌkænəlaɪˈzeɪʃn/
 n. 运河网;开运河
capillary /kəˈpɪləri/
 adj. 毛细管的;毛状的
 n. 毛细管
cavitation /ˌkævɪˈteɪʃn/
 n. 空化;汽蚀;空洞形成,气穴现象;涡凹

cistern /ˈsistən/
  n. 水箱；水池；贮水器
Cretaceous /kriˈteiʃəs/
  adj. 白垩纪的；似白垩的
  n. 白垩纪；白垩系
debris /ˈdebriː/
  n. 碎片，残骸；残渣；岩屑
deplete /diˈpliːt/
  vt. 耗尽，用尽；使衰竭，使空虚
depositional /ˌdepəˈziʃənl/
  adj. 沉积作用的
derris /ˈderis/
  n. 〈植〉鱼藤；鱼藤属
diameter /daiˈæmitə(r)/
  n. 直径，直径长；放大率
disperse /diˈspəːs/
  vt. 分散；使散开；传播
  vi. 分散
  adj. 分散的
eon /ˈiːən/
  n. 永世；无数的年代；极长时期
exploitation /ˌeksplɔiˈteiʃn/
  n. 开发，开采；利用；广告推销；剥削
exploration /ˌekspləˈreiʃn/
  n. 探测；探究；踏勘
fallow /ˈfæləu/
  vt. 使(土地)休闲；潜伏
flourish /ˈflʌriʃ/
  n. 兴旺；茂盛
  vi. 繁荣，兴旺；茂盛；活跃；处于旺盛时期
foraminifera /fəˌræməˈnifərə/
  n. 有孔虫
gastropod /ˈgæstrəpɔd/
  n. 腹足类动物
generator /ˈdʒenəreitə(r)/
  n. 发电机，发生器；电力公司；生产者，创始者
grid /grid/
  n. 格子；地图上的坐标方格；(输电线路、天然气管道等的)系统网络

hydrocarbon /ˌhaidrəˈkɑːbən/
  n. 〈化〉碳氢化合物
ignite /igˈnait/
  vt. 点燃；使燃烧；使激动
  vi. 点火
impoverishment /imˈpɔvəriʃmənt/
  n. 贫穷；致使贫乏或贫瘠之事物
improvise /ˈimprəvaiz/
  vt. 即兴创作；即兴表演；临时做；临时提供
impurity /imˈpjuəriti/
  n. 杂质；不纯；不洁
indigenous /inˈdidʒinəs/
  adj. 本土的；土著的
jointly /ˈdʒɔintli/
  adv. 共同地，联合地，连带地
limestone /ˈlaimstəun/
  n. 〈岩〉石灰岩
megawatt /ˈmegəwɔt/
  n. 兆瓦，百万瓦特（电能计量单位）
molecular /məˈlekjulə(r)/
  adj. 〈化〉分子的；由分子组成的
mollusk /ˈmɔləsk/
  n. 〈动〉软体动物
mound /maund/
  n. 堆；高地；坟堆；护堤
mound-builder
  n. 筑墩人
nitrogen /ˈnaitrədʒən/
  n. 〈化〉氮
panhandle /ˈpænhændl/
  n. 平锅柄；〈美〉柄状的狭长区域
  vt. 向……乞讨
  vi. 乞讨
penetrate /ˈpenitreit/
  vt. 渗透；穿透；洞察
  vi. 渗透；刺入；看透
permeable /ˈpəːmiəbl/
  adj. 能透过的；有渗透性的

pore /pɔː/
 vi. 细想；凝视
 n. 气孔；小孔
 vt. 使注视

porous /ˈpɔːrəs/
 adj. 多孔渗水的；能渗透的；有气孔的

profusion /prəˈfjuːʒn/
 n. 丰富，充沛

ramp /ræmp/
 n. 斜坡，坡道；敲诈

reef /riːf/
 n. 暗礁；〈地质〉矿脉；收帆

reservoir /ˈrezəvwɑː(r)/
 n. 〈水利〉水库；油箱

robust /rəʊˈbʌst/
 adj. 精力充沛的；坚定的

saline /ˈseɪlaɪn/
 adj. 含盐的，咸的
 n. 盐湖，盐泉；盐溶液

scourge /skɜːdʒ/
 n. 鞭；灾祸

seal /siːl/
 n. 密封；印章；海豹；封条；标志
 vt. 密封；盖章

seaweed /ˈsiːwiːd/
 n. 〈植〉海草，海藻

stratigraphic /ˌstrætɪˈgræfɪk/
 adj. 地层的；地层学的

stubborn /ˈstʌbən/
 adj. 顽固的；顽强的；难处理的

submerge /səbˈmɜːdʒ/
 v. 淹没；把……浸入；沉没，下潜；使沉浸

suspend /səˈspend/
 vt. 延缓，推迟；使暂停；使悬浮

tidal /ˈtaɪdl/
 adj. 潮汐的；潮水的；由于潮水作用的；定时涨落的

transgression /trænsˈgreʃn/
 n. 〈地质〉海侵；犯罪；违反；逸出

turbine /ˈtɜːbaɪn/
 n. 汽轮机；涡轮机；透平机

unethical /ʌnˈeθɪkl/
 adj. 不道德的；缺乏职业道德的

undercut /ˌʌndəˈkʌt/
 v. （网球等）从下削球；廉价出售；较便宜的工资工作

water lettuce
 n. 水浮莲

Unit 8

adjacent /əˈdʒesənt/
 adj. 相邻；邻近的，毗邻的

aftermath /ˈæftəmæθ/
 n. 后果；余波

amplitude /ˈæmplɪtuːd/
 n. 振幅；广阔；角度距离

basin /ˈbeɪsn/
 n. 盆地；流域

brittle /ˈbrɪtl/
 adj. 易碎的

calving /ˈkɑːvɪŋ/
 n. 裂冰（作用）
 v. （冰川）崩解

compress /kəmˈpres/
 vt. 压缩；压紧

conductive /kənˈdʌktɪv/
 adj. 传导的

convective /kənˈvektɪv/
 adj. 传送性的，对流的

convergent /kənˈvɜːdʒənt/
 adj. 趋同的；会合的；逐渐减小的

displacement /dɪsˈplesmənt/
 n. 取代，替代

divergent /dɪˈvɜːdʒənt; daɪ-/
 adj. 发散的；扩散的

elastic /ɪˈlæstɪk/
 adj. 有弹力的；可伸缩的

epicentre /ˈepisentə(r)/
  n. 震中；中心；集中点
equilibrium /ˌi:kwiˈlibriəm/
  n. 平衡，均势；平静
feature /ˈfi:tʃə(r)/
  n. 特征，特点
  vt. 使有特色
flexure /ˈflekʃə(r)/
  n. 屈曲，弯曲部分，打褶
fracture /ˈfræktʃə(r)/
  vt. 折断，破碎
glacier /ˈglæsiə/
  n. 冰河，冰川
gravitational /ˌgræviˈteiʃənl/
  adj. 万有引力的，重力的
hypocentre /ˌhaipəuˈsentə(r)/
  n. 震源
inundate /ˈinʌndeit/
  vt. 淹没；泛滥
negligible /ˈneglidʒəb(ə)l/
  adj. 微不足道的；可以忽略的
oblique /əˈblik/
  adj. 倾斜的；间接的
overlying /əuvəˈlaiiŋ/
  v. 叠加；躺在……上面
perturbation /ˌpə:rtərˈbeiʃn/
  n. 忧虑；不安；烦恼；摄动
premium /ˈpri:miəm/
  n. 保险费
propagation /ˌprɔpəˈgeiʃn/
  n. 传导
regime /reiˈʒi:m/
  n. 方法
rupture /ˈrʌptʃə(r)/
  n. 断裂，破裂
  vt. 使破裂
seismicity /saizˈmisiti/
  n. 地震活动；受震强度
thirtyfold /ˈθə:tifəuld/
  adj. 三十倍的
  adv. 三十倍地
thrust /θrʌst/
  n. 推力；逆断层
tsunami /tsuˈnɑ:mi/
  n. 海啸
wavelength /ˈweivleŋkθ/
  n. 波段；波长

Keys

Unit 1
Introduction to Earth Science

Passage A

I. 1. Geology is the science that pursues an understanding of planet Earth. (Para. A)
2. Physical geology and historical geology. (Para. A)
3. Firstly, physical geology examines the materials composing Earth and seeks to understand the many processes that operate beneath and upon its surface. The aim of historical geology, on the other hand, is to understand the origin of Earth and its development through time. It strives to establish a chronological arrangement of the multitude of physical and biological changes that have occurred in the geologic past.
   Secondly, the study of physical geology logically precedes the study of Earth history because we must first understand how Earth works before we attempt to unravel its past. (Para. A)
4. Volcanoes, floods, earthquakes, and landslides. (Para. D)
5. Geologic studies contribute to our understanding of natural hazards and the impacts of human activities on Earth. They also deal with the formation and occurrence of vital resources, and maintaining supplies and the environmental impact of their extraction and use. (Para. D—F)
6. Aristotle believed that rocks were created under the influence of the stars and that earthquakes occurred when air in the ground was heated by central fires and escaped explosively. For example, he attributed the finding of a fossil fish to the reason that "a great many fishes live on Earth motionless and are found when excavations are made".
   Though his explanations were inadequate, they continued to be expounded for many centuries, thus thwarting the acceptance of more up-to-date ideas. (Para. G)
7. Briefly stated, catastrophists believed that Earth's landscapes had been shaped primarily by great catastrophes. Features such as mountains and canyons, which today we know take great periods of time to form, were explained as having been produced by sudden and often worldwide disasters caused by unknown forces that were no longer in operation. (Para. H)

Uniformitarianism suggests that the physical, chemical, and biological laws that operate today also operated in the geologic past. In other words, the forces and processes that we observe shaping our planet today have been at work for a very long time. Thus, to understand ancient rocks, we must first understand present-day processes and their results. (Para. I)

8. Catastrophists believed that Earth's landscapes had been shaped primarily by great catastrophes.

Uniformitarians believe that geological processes can continue over extremely long periods of time.

Therefore, the former believe that the Features such as mountains and canyons on Earth were produced by sudden and often worldwide disasters caused by unknown forces that were no longer in operation, whereas the latter believe they take great periods of time to form. (Para. J and K)

II. 1. landslides 2. vulnerable
3. hazards 4. confronted
5. impact 6. erosion/eroded
7. adversely 8. composed
9. chronological 10. unravel

III. 1. The particles in liquids vibrate vigorously. As a result, some particles can gain sufficient energy to escape the liquid.
2. Rocks on level areas are likely to remain in place over time, whereas the same rocks on slopes tend to move as a result of gravity.
3. If a well is not drilled deep enough, it may stop producing water in periods of prolonged drought. On the other hand, money is wasted if the well is drilled beyond where the well could conceivably ever go dry.
4. All of the planet's forests account for approximately three-quarters of all the biomass on one-fifth of the land. In contrast, deserts cover about the same land area but account for less than 2% of the biomass.
5. Hurricanes can form under appropriate conditions all over the world, but are named differently in different oceans; for example, they are called typhoons in the Pacific Ocean and cyclones in the Indian Ocean.

IV. 1. 相反,地史学的目的是了解地球的起源以及它的发展历史。因此,它主要致力于对地质史上曾经发生过的诸多物理或生物现象进行时间排序。
2. 按道理讲,普通地质学先于地球史学,因为我们必须先了解地球是如何运转的,才有可能了解它过去是什么样的。
3. 另有一些大城市,如果土地使用不当或建筑不合格的话,面临地震和火山爆发的威胁。尤其是人口的飞速增长使得其城市体系更显脆弱。
4. 有些地貌如高山和峡谷,当然我们现在知道它们的形成需要漫长的时间。当时却被解释为是突然性的、全球性的灾难造成的,而且造成这些灾难的外力迄今还不清楚,也不复存在了。

5. 有一点很重要,就是现今的很多地貌特征就我们所能够观察得到的这几十年里似乎一直没什么变化,其实它们一直在变化着,只是这些变化是相对于几百年、几千年甚至是好几百万年来说的。

Ⅴ. Sir Charles Lyell, James Hutton, Alfred Wegener and Harry Hess are all famous geologists in history. They helped form the theories, ideas and investigations of Plates Tectonics, Sea Floor-spreading and the development of the theory uniformitarianism over catastrophism. Among them, James Hutton was born in Edinborough, Scotland on June 3, 1726, and died on March 26, 1797. He is known as the "founding father of modern geology". He started the theory of uniformitarianism, which directly influenced Sir Charles Lyell and Charles Darwin later.

Passage B

Questions 1—4: 1. xii   2. x   3. i   4. iv
Questions 5—10: 5. Y   6. NG   7. N   8. Y   9. N   10. NG
Questions 11—13: 11. gaseous   12. spheres   13. integrate

Unit 2
Earth History and Life Evolution

Passage A

Ⅰ. 1. Relative dating to determine the age of rocks and fossils; determining the numerical age of rocks and fossils; using palaeomagnetism to date rocks and fossils.
2. the principle of original horizontality; the principle of superposition; the principle of cross-cutting relationships; the principle of faunal succession
3. Unlike relative dating methods, absolute dating methods provide chronological estimates of the age of certain geological materials associated with fossils, and even direct age measurements of the fossil material itself.
4. Geologists commonly use radiometric dating methods, based on the natural radioactive decay of certain elements such as potassium and carbon, as reliable clocks to date ancient events.
5. Electron spin resonance and thermo luminescence.
6. When rocks or fossils become older than 100 000 years.
7. Earth is like a gigantic magnet. It has a magnetic north and south pole and its magnetic field is everywhere. Just as the magnetic needle in a compass will point toward magnetic north, small magnetic minerals that occur naturally in rocks point toward magnetic north, approximately parallel to Earth's magnetic field.

8. Geologists can measure the palaeomagnetism of rocks at a site to reveal its record of ancient magnetic reversals. Every reversal looks the same in the rock record, so other lines of evidence are needed to correlate the site to the GPTS. Information such as index fossils or radiometric dates can be used to correlate a particular palaeomagnetic reversal to a known reversal in the GPTS. Once one reversal has been related to the GPTS, the numerical age of the entire sequence can be determined.

II. 1. abundance 2. dislodge from
3. lava 4. elapsed
5. gigantic 6. incorporate into
7. chronological 8. deposit
9. fossil 10. reversal

III. 1. Once he has made up his mind to choose the course, he will not be swerved from his course.
2. Once you find mistake, you should correct.
3. Once the environmental damage is done, it takes many years for the ecosystem to recover.
4. Once virtue is lost, all is lost.
5. Once you understand this rule, you will have no further difficulty.

IV. 1. 地质学家已经确立了一套适用于裸露于地球表面的沉积岩和火山岩的原则，可以据此确定岩石记录中地质事件的相对年龄。
2. 地质学家通常使用放射性测年法，即将某些元素（如钾和碳）的自然放射性衰变作为测算古代事件的可靠计时方式。
3. 诸如电子自旋共振和热发光等其他方法可以评估放射性对矿物晶体结构中缺陷（或"陷阱"）中电子积累的影响，地质学家还用这些方法来确定岩石或化石的年龄。
4. 辐射是放射性衰变的副产品，它使电子从原子中的正常位置移开，并陷入材料晶体结构缺陷之中。
5. 地质学家可在某一地点测量岩石的古地磁，从而获得其古代磁性反转的记录。

V. The study on fossils enables one to understand the evolution of biology, and is helpful in determining the age of the strata. One of the methods used to determine fossil geochronology is the relative dating method, which is based on the principle that new strata are deposited on older strata. Therefore, the relative ages of the strata can be determined by their location in the sequence if they are not disturbed by the fault or other factors. In addition, the relative age can be measured according to the biological evolution. In general, the newer the strata are, the more advanced and more complex the living organisms are. Meanwhile, diverse types of fossils and their assemblages are contained in strata of different ages whereas strata formed at the same time and under the same geographical environment have the same fossil or fossil assemblage no matter how far apart they are.

Passage B

Questions 1—6: 1. iii 2. i 3. ii 4. vi 5. v 6. iv

Questions 7—9: 7. B   8. D   9. C
Questions 10 and 11: 10. B   11. D
Questions 12 and 13: 12. B   13. C

Unit 3
Plate Tectonics and Earth Structure

Passage A

Ⅰ. 1. The puzzlelike fit of the continents, particularly South America and Africa. (Para. A)

2. It challenges the long-held assumption that the continents and ocean basins had fixed geographic positions. On the contrary, it hypothesised that about 200 million years ago, during the early part of the Mesozoic era, this supercontinent began to fragment into smaller landmasses. These continental blocks then "drifted" to their present positions over a span of millions of years. (Para. B)

3. Firstly, the puzzlelike fit of the continents. Secondly, fossils of the reptile Mesosaurus found in South America match across the seas with the fossils found in Africa. (Para. C) Thirdly, rock types and geologic features such as mountain belts were found in a particular region on one continent closely match in age and type of those found in adjacent positions on the once adjoining continent. (Para. D)
Lastly, evidence for a glacial period that dated to the late Palaeozoic had been discovered in southern Africa, South America, Australia, and India. This meant that about 300 million years ago, these areas were closely collected. (Para. E)

4. Because their fossils were found in many areas which indicate these areas might be together in the history.

5. The rocks found in a particular region on one continent closely match in age and type of those found in adjacent positions on other continents, which indicate these continents might be connected in history and drifted apart later on. (Para. D)

6. Wegener suggested that these continents were connected and partly covered with ice near Earth's south pole long ago. This would account for the conditions necessary to generate extensive expanses of glacial ice over much of these landmasses. (Para. E)

7. One main objection to Wegener's hypothesis stemmed from his inability to identify a credible mechanism for continental drift. Wegener also incorrectly suggested that the larger and sturdier continents broke through thinner oceanic crust, much like ice breakers cut through ice. However, no evidence existed to suggest that the ocean floor was weak enough to permit passage of the continents without the continents being appreciably deformed in the process.

Ⅱ. 1. landmass   2. considerable

3. opposed 4. convince
5. fossil 6. asserted
7. evidence 8. supporting
9. hypothesised 10. deposits

Ⅲ. 1. One of the major tenets of the continental drift hypothesis was that a supercontinent called Pangaea began breaking apart about 200 million years ago.

2. Modern maps of the seafloor substantiate Wegener's contention that, if land bridges of this magnitude once existed, their remains would still lie below sea level.

3. Ol Doinyo Lengai is an active volcano in the East African Rift Valley, a place where Earth's crust is being pulled apart.

4. Wegener was unable to explain exactly, how the continents drifted apart.

5. To add credibility to his argument, Wegener documented cases of several fossil organisms that were found on different landmasses despite the unlikely possibility that their living forms could have crossed the vast ocean presently separating them.

Ⅳ. 1. 大陆,尤其是南美洲和非洲,二者跟拼图似的可以合在一起,这种提法出现在17世纪,即比较完善的世界地图出现的年代。

2. 如果这些大陆曾经是合在一起的,那么在一个大陆某个特定位置发现的岩石应该在年代和种类上跟曾经是相邻大陆的相近位置发现的石头吻合。

3. 比如,包括阿巴拉契亚山脉在内的造山带向东南方向一直延展至美国东部,然后在纽芬兰海岸消失。

4. 大家敬仰的美国地质学家洛林·托马斯·张伯伦甚至如此评论,"大体说来,魏格纳的假设是非常不严谨的,它在怎样看待我们的地球问题上,要么想法过于无拘无束,没有考虑其局限性,要么建立在牵强附会或有问题的事实依据基础上,完全没有办法跟其他同类理论相比。它的整个论述就好比在玩一个没有游戏规则、没有明确规范的游戏。"

5. 随着科技的进步,以及大陆漂移说观点的不断发展,更多关于大陆漂移说的证据在魏格纳去世后被找到,并最终促成了板块构造说理论的诞生。

Ⅴ. In 1915, the German geologist and meteorologist Alfred Wegener first proposed the theory of continental drift. Wegener hypothesised that there was a gigantic supercontinent 200 million years ago, which he named Pangaea, meaning "All-land". Pangaea started to break up into two smaller supercontinents, called Laurasia and Gondwanaland, during the Jurassic period. By the end of the Cretaceous period, the continents were separating into landmasses that look like our modern-day continents. Wegener published this theory in his 1915 book, *On the Origin of Continents and Oceans*.

Passage B

Questions 1—4: 1. xiii 2. xi 3. iv 4. i
Questions 5—10: 5. Y 6. NG 7. Y 8. N 9. NG 10. N
Questions 11—13: 11. E 12. H 13. B

Unit 4
Mineral and Rock

Passage A

I. 1. A mineral has one specific chemical composition, whereas a rock can be an aggregate of different minerals or mineraloids.
   2. A mineral is an element or chemical compound that is normally crystalline and that has been formed as a result of geological processes. In addition, biogenic substances were explicitly excluded. Biogenic substances are chemical compounds produced entirely by biological processes without a geological component (e. g., urinary calculi, oxalate crystals in plant tissues, etc.) and are not regarded as minerals. However, if geological processes were involved in the genesis of the compound, then the product can be accepted as a mineral.
   3. The requirement of an ordered atomic arrangement is usually synonymous with crystallinity; however, crystals are also periodic, so the broader criterion is used instead. An ordered atomic arrangement gives rise to a variety of macroscopic physical properties, such as crystal structure and habit, hardness, luster, color and streak.
   4. In other cases, minerals can only be classified by more complex optical, chemical or X-ray diffraction analysis; these methods, however, can be costly and time-consuming.
   5. An example of this property exists in kyanite, which has a Mohs hardness of 5½ parallel to [001] but 7 parallel to [100].
   6. Diamond is the hardest natural material.
   7. An example of this property exists in kyanite, which has a Mohs hardness of 5½ parallel to [001] but 7 parallel to [100]. 7 Examples of minerals with this lustre are galena and pyrite.
   8. The diaphaneity of a mineral depends on thickness of the sample.

II. 1. aggregate   2. encompassed
    3. composition   4. criteria
    5. excluded   6. synonymous
    7. explicitly   8. underlying
    9. stability   10. Periodic

III. 1. Pensions are linked to inflation, whereas they should be linked to the cost of living.
     2. You eat a massive plate of food for lunch, whereas I have just a sandwich.
     3. He had never done anything for them, whereas they had done everything for him.
     4. One's life is finite, whereas one's passion for life is infinite.
     5. These gases trap Sun's heat whereas sulphur dioxide cools the atmosphere.

Ⅳ. 1. 矿物是一种天然的化学化合物,通常是结晶形式和无机成因。

2. 第一个标准是一种矿物必须由自然过程形成,它不包括人为化合物。

3. 有序的原子排列产生了各种宏观物理性质,如晶体结构和习性、硬度、光泽、颜色和条纹。

4. 一种矿物可以通过多种物理性质来鉴定,有些是充分鉴定而不含糊其辞的。

5. 光泽指光如何从矿物表面反射,与它的质量和强度有关。

Ⅴ. China has discovered 171 varieties of minerals, and 158 of them with proved reserves. There are ten energy-related minerals, including oil, natural gas, coal, uranium and geotherm; 54 metallic minerals, including iron, manganese, copper, aluminum, lead and zinc; 91 non-metallic minerals, including graphite, phosphorus, sulfur and sylvite; and Three liquid minerals, including groundwater, and mineral water. There are nearly 18 000 mineral deposits in China, including more than 7 000 big and medium-sized ones.

Passage B

Questions 1—4: 1. ii   2. i   3. v   4. iv

Questions 5—10: 5. N   6. Y   7. N   8. Y   9. Y   10. NG

Questions 11—13: 11. F   12. E   13. H

Unit 5
Geomorphology and Geography

Ⅰ. 1. Weathering refers to the group of destructive processes that change the physical and chemical character of rock on or near Earth's surface.

2. Rocks and soils can be broken down through direct contact with atmospheric conditions such as heat, water, ice and pressure.

3. Chemical weathering changes the materials that make up rocks and soil.

4. There are 3 agents of erosion: water, wind and ice.

5. Waves constantly crash against shores. They pound rocks into pebbles and reduce pebbles to sand. Water sometimes takes sand away from beaches. This moves the coastline farther inland.

6. Wind is responsible for some sand dunes in some area of the Gobi Desert. It carries dust, sand, and volcanic ash from one place to another. Wind can sometimes blow sand into towering dunes.

7. In frigid areas and on some mountaintops, glaciers move slowly downhill and across the land. As they move, they pick up everything in their path, from tiny grains of sand to huge boulders. The rocks carried by a glacier rub against the ground below, eroding both the ground and the rocks. Glaciers grind up rocks and scape away the soil. Moving glaciers gouge out basins and from steep-sided mountain valleys.

8. Weathering breaks down rocks that are either stationary or moving. Erosion is the picking

Keys

up or physical removal of rock particles by an agent such as streams, wind or glaciers. Weathering helps break down a solid rock into loose particles that are easily eroded. Most eroded rock particles are at least partially weathered, but rock can be eroded before it had weathered at all. A stream can erode weathered or unweathered rock fragments.

II. 1. break down 2. frigid
3. deterioration 4. wearing away
5. take place 6. towering
7. incorporated 8. fragments
9. decomposition 10. alters

III. 1. Kate saw her brother Bill as she was getting off the school bus.
2. Mother dropped her glass to the ground as she was standing up from her seat.
3. As the thief was bolting out of the house, a policeman fired at him.
4. Sue thought of her talk with her mother as she ran to catch the school bus.
5. As the teacher entered the classroom, all the students shouted "Happy Birthday" to him.

IV. 1. 风化作用是一种缓慢但有效的作用力,即使最坚硬的岩石也易受到影响。
2. 受到风化变弱的岩石更容易侵蚀,这个过程是重力、流水、风或冰搬运岩块,并将其沉积到其他地方。
3. 当受热时,岩石中的不同矿物膨胀程度不一,例如,在相同的受热条件下,石英颗粒的膨胀量是斜长岩颗粒膨胀量的3倍。
4. 几乎所有的岩石都有一些裂纹和空洞。植物和树木可以在表部岩石中的裂隙生根。
5. 因为水将离子携带到反应发生的地方,并参与反应,然后将反应产物带着,因此水是控制化学风化速率唯一的最重要因素。

V. Weathering plays a vital role in our daily lives, with both positive and negative outcomes. It free life-sustaining minerals and elements from solid rock, allowing them to become incorporated into our soils and finally into our foods. Indeed, we would have very little food without weathering, as this process produces the very soil in which much of our food is grown. But weathering can also weak havoc on the structures we build. Countless monuments—from the pyramids of Egypt to ordinary tombstones—have suffered drastic deterioration from freezing water, hot sunshine, and other climatic forces.

Passage B

Questions 1—5: 1. iii 2. i 3. iv 4. ii 5. vii
Questions 6—11: 6. Y 7. NG 8. N 9. Y 10. NG 11. Y
Questions 12—14: 12. F 13. A 14. B

Unit 6
Climate Change and Atmosphere

Passage A

Ⅰ. 1. Climate means the average weather.

2. The classical time period is 30 years.

3. Natural causes, e.g., changes in Sun's output, or human activities, e.g., changing the composition of the atmosphere.

4. Climate change reflects a change in the energy balance of the climate system, i.e., changes the relative balance between incoming solar radiation and outgoing infrared radiation from Earth.

5. Changes in the energy balance of the climate system.

6. Two types: internal forcing mechanism, and external forcing mechanism.

7. "Global warming" refers to the change in Earth's global average surface temperature.

8. Global warming is closely associated with a broad spectrum of other climate changes, such as: increases in the frequency of intense rainfall, decreases in snow cover and sea ice, more frequent and intense heat waves, rising sea levels, and widespread ocean acidification.

9. To limit the magnitude of changes to the climate and the impacts on communities and wildlife.

Ⅱ. 1. anticipated 2. induce
3. variation 4. tributary
5. elaborate 6. abruptly
7. internal 8. devastate
9. spectrum 10. curb

Ⅲ. 1. Regardless who your stakeholders are, get out there with them and see first-hand how they benefit from and/or struggle with your software.

2. And since all cloud systems perform under the same general concepts, this technique should be useful to you regardless which cloud platform (or platforms) you choose to employ.

3. So we have a unanimous vote for that candidate regardless how they finished in the Iowa caucuses.

4. To accurately reproduce the narrative in the output document, the style sheet must handle elements regardless where they appear, and the push model excels at that.

5. Regardless whether crime or misfortune was predominant in any given story, the faces in that world were specific and personal.

Ⅳ. 1. 气候变化这个短语用来描述气候的变化,根据其统计特性,例如全球平均表面温度来衡量。

2. 所指的气候变化可能是由于自然原因,例如,太阳输出的变化或人类活动引起的,例如改变大气的组成。

3. 气候变化反映了气候系统能量平衡的变化,即改变地球的入射太阳辐射和向外辐射之间的相对平衡。

4. 气候变化最一般的定义是指不管原因是什么,在气候系统的长期变化中考虑统计特性,而全球变暖指的是地球全球平均表面温度的变化。

5. 国家野生动物联合会说,为了限制气候变化的大小和对社区和野生动物的影响,我们必须控制导致全球变暖的污染。

Ⅴ. Today we envision Earth as dynamic, ever-changing, and evolving under the influences of many complex physical, chemical, and biological processes that show great variations of rates and intensities. These processes complete with one another to produce a situation tending toward some balance, or equilibrium, among them. But the equilibrium is always being disturbed by new changes. The resulting modifications are in part nonrepeating changes, and in part repeating ones. Thus, American geologist C. R. van Hies reflected this newer view of Earth in 1898 when he wrote that: "Earth is not finished, but is now being, and will forevermore be re-made."

Passage B

Question 1—4: 1. E   2. AB   3. C   4. D
Question 5—9: 5. F   6. F   7. NG   8. T   9. T
Question 10—13: 10. constant   11. Orbit   12. instabilities   13. cycles   14. random

Unit 7
Natural Resources and Environment Protection

Passage A

Ⅰ. 1. It is many millions of years.

2. Source rocks are the rocks, such as sandstone, shale, or limestone.

3. Pool is a term that has special meaning in the oil industry; it refers to an economically producible quantity of oil dispersed in rock within Earth.

4. Rock strata that contain economically recoverable concentrations of oil is called reservoirs.

5. They aree vaporitic salts, faulting, and stratigraphic traps.

6. No, not all of the gas and liquids would be driven out, no matter how great the driving pressure is.

7. Several techniques have been devised to increase the yield of oil from reservoirs, such as water, steam, or gas injection, and even igniting some of the oil.

8. 30%—40%.

Ⅱ. 1. is sealed 2. profusion
3. transgression of 4. disperse
5. suspend 6. penetrate
7. flourish 8. ignite
9. fault 10. pore

Ⅲ. 1. Contrary to the common description of the cold war as a conflict between east and west, communism was a prototypically western ideology, with roots in the European enlightenment.
2. Contrary to those people, many others think advertisements are very unpleasant. Consumers are often cheated by the false advertisement on which consumers always waste a great deal of time.
3. It also shows that agriculture in developed countries will not be decimated, contrary to claims of vested farm interests in those countries.
4. There's an upward bias to the stock market, contrary to what you may have seen recently.
5. We can't be everywhere at once, and contrary to popular belief, you cannot always do everything you want to do.

Ⅳ. 1. 科学家们认为，石油的形成始于数百万年前，那时低等的动植物在海洋中繁盛生长，就像今天这样。
2. 与人们普遍认为的相反，石油不会出现在地下，而是储存在类似蓄水池的"水池"里，可以被挖掘和泵干。
3. 在油气形成和积聚过程中，一些原始海水被移位，重力将气体、油和水分离成层，但实际上油藏内部的情况要复杂得多。
4. 然而，不管驱动压力有多大，并不是所有的气体和液体都会被排出。
5. 如果一口井穿过这个区域，压力就会把岩石中的油、气和盐水驱出井里。

Ⅴ. Some of these works are based on study of newly released industrial data sets, while others are purely of academic origin. Taken together they represent a significant advance in our understanding of the region and help lay the groundwork for the upcoming scientific drilling in the deep-water part of the South China Sea in February－March 2014 by the International Ocean Discovery Program (IODP). Although there remains significant debate about how strain was accommodated during opening, it is clear that the South China Sea was generated as a result of extension of relatively warm continental crust that had been influenced by an earlier phase of subduction under southern China known as the Cathaysia arc.

Passage B

Questions 1－4: 1. C 2. E 3. A 4. C
Questions 5－9: 5. A 6. D 7. F 8. F 9. J
Questions 10－14: 10. maintenance 11. slow 12. low pressure 13. cavitation

Unit 8
Geological Hazard

Passage A

I. 1. Earthquakes, volcanic eruptions and other underwater explosions (including detonations of underwater nuclear devices), landslides, glacier calvings, meteorite impacts and other disturbances above or below water all have the potential to generate a tsunami.
2. Tsunami waves do not resemble normal undersea currents or sea waves, because their wavelength is far longer.
3. Rather than appearing as a breaking wave, a tsunami may instead initially resemble a rapidly rising tide, and for this reason they are often referred to as tidal waves, although this usage is not favoured by the scientific community because tsunamis are not tidal in nature.
4. The 2004 Indian Ocean tsunami was among the deadliest natural disasters in human history with at least 230 000 people killed or missing in 14 countries bordering the Indian Ocean.
5. The term seismic sea wave also is used to refer to the phenomenon, because the waves most often are generated by seismic activity such as earthquakes. seismic sea wave is not a completely accurate term, as forces other than earthquakes—including underwater landslides, volcanic eruptions, underwater explosions, land or ice slumping into the ocean and the weather when the atmospheric pressure changes very rapidly—can generate such waves by displacing water.
6. The principal generation mechanism or cause of a tsunami is the displacement of a substantial volume of water or perturbation of the sea.
7. More specifically, a tsunami can be generated when thrust faults associated with convergent or destructive plate boundaries move abruptly, resulting in water displacement, owing to the vertical component of movement involved.
8. Tsunamis cause damage by two mechanisms: the smashing force of a wall of water travelling at high speed, and the destructive power of a large volume of water draining off the land and carrying a large amount of debris with it, even with waves that do not appear to be large.

II. 1. landslides   2. tectonic
    3. debris   4. devastated
    5. compresses   6. adjacent
    7. seismic   8. wavelengths
    9. amplitudes   10. shallow

Ⅲ. 1. It exists in dreams rather than actuality.
2. They aim at quality rather than quantity.
3. Boxing is a test of skill and technique, rather than brute strength.
4. Let's get it over with as soon as possible, rather than drag it out.
5. You are buying direct, rather than through an agent.

Ⅳ. 1. 地震、火山喷发、山体滑坡、冰川崩裂以及水面上或者水面下的其他扰动都有可能引发海啸。
2. 海啸波不像普通的海底洋流或海浪，因为它们的波长要长得多。
3. 2004年的印度洋海啸是人类历史上最严重的自然灾害之一，在印度洋沿岸的14个国家中至少有23万人死亡或失踪。
4. 在英语国家使用"海啸"这个词之前，科学家们通常建议使用"地震波"，而不是"海浪波"。
5. 海啸造成的破坏有两种机制：一种是以高速行进的水墙产生巨大的冲击力，另一种是即使海浪看起来并不大，但是大量的水从陆地上引流产生巨大的破坏力并且携带大量的碎片。

Ⅴ. The 2004 Indian Ocean earthquake, known by the scientific community as the Sumatra-Andaman earthquake, was an undersea earthquake that occurred at 00:58:55 on December 26, 2004. The earthquake generated a tsunami that was among the deadliest disasters in modern history, killing well over 200 000 people. Various values were given for the magnitude of the earthquake, ranging from 9.0 to 9.3 (which would make it the second largest earthquake ever recorded on a seismograph since 1900), though authoritative estimates now put the magnitude at 9.15. In May 2005, scientists reported that the earthquake itself lasted nearly ten minutes when most major earthquakes last no more than a few seconds.

Passage B

Questions 1—4: 1. v   2. ii   3. vi   4. iv
Questions 5—10: 5. N   6. Y   7. N   8. Y   9. N   10. NG
Questions 11—13: 11. normal   12. increase   13. hierarchy

图书在版编目(CIP)数据

地学英语阅读教程/赖小春,汪卫红主编．—武汉:中国地质大学出版社,2018.2
地学英语阅读系列教材/赖旭龙,董元兴,刘芳,赖小春主编
ISBN 978-7-5625-4234-6

Ⅰ.①地⋯

Ⅱ.①赖⋯ ②汪⋯

Ⅲ.①地质-英语-阅读教学-教材

Ⅳ.①P5

中国版本图书馆 CIP 数据核字(2018)第 028000 号

地学英语阅读教程			赖小春 汪卫红	主 编
			王晓婧 付 蕾 曾艳霞 胡冬梅 刘雪莲	副主编
责任编辑:龙昭月 谌福兴	选题策划:李国昌 张 琰		责任校对:张咏梅	
出版发行:中国地质大学出版社(武汉市洪山区鲁磨路388号)			邮政编码:430074	
电 话:(027)67883511	传 真:(027)67883580		E-mail:cbb@cug.edu.cn	
经 销:全国新华书店			http://cugp.cug.edu.cn	
开本:850 毫米×1168 毫米 1/16		字数:275 千字		印张:9.75
版次:2018 年 2 月第 1 版		印次:2018 年 2 月第 1 次印刷		
印刷:武汉市籍缘印刷厂		印数:1—1000 册		
ISBN 978-7-5625-4234-6				定价:38.00 元

如有印装质量问题请与印刷厂联系调换